WASPS

OF THE WORLD

WASPS
OF THE WORLD

A GUIDE TO
EVERY FAMILY

Simon van Noort
and Gavin Broad

PRINCETON UNIVERSITY PRESS
PRINCETON AND OXFORD

Published in 2024 by Princeton University Press
41 William Street, Princeton, New Jersey 08540
99 Banbury Road, Oxford OX2 6JX
press.princeton.edu

Library of Congress Control Number 2023934865
ISBN: 978-0-691-23854-8
Ebook ISBN: 978-0-691-25705-1

Printed and bound in Malaysia
10 9 8 7 6 5 4 3 2 1

This book was conceived, designed, and produced by
The Bright Press, an imprint of the Quarto Group
1 Triptych Place, 2nd Floor, London SE1 9SH, United Kingdom
www.Quarto.com

Publisher **James Evans**
Editorial Director **Isheeta Mustafi**
Managing Editor **Jacqui Sayers**
Art Director and Cover Design **James Lawrence**
Senior Editor **Joanna Bentley**
Project Manager **David Price-Goodfellow**
Design **Wayne Blades**
Illustrations **John Woodcock**
Picture Research **Jane Smith**

Cover photos: Front, clockwise from top left, all Shutterstock: Silmiart,
Alex Puddephatt, Ed Phillips, khlungcenter, Vitalii Hulai, Melinda
Fawver, Erik Karits, Huw Penson, ASGOLD, alslutsky, Vitalii Hulai,
Abu Zakaria. Spine: Shutterstock/Kletr. Back cover: Shutterstock/
Roberto Michel.

CONTENTS

INTRODUCTION

WHAT IS A WASP?

That black and yellow insect with a fearsome sting and a cinched waist is a wasp, and most of us would recognize it as such, but that waspy image is completely inadequate to the task of describing an almost overwhelming world of wasps with a diversity of shapes and colors. These are the insects that dominate much of what we see around us: the insect order Hymenoptera (wasps, bees, and ants). Certain groups of Hymenoptera have acquired widely used vernaculars, such as ants and bees, but the vast majority of Hymenoptera are called "wasps" of some sort or other, such as cuckoo wasps, ensign wasps, woodwasps, etc. Although ants and bees are often readily distinguished by the lay person, both groups are in fact derived wasps closely related to other wasp families, each having independently evolved from within the wasp tree of life.

Hence, wasps cannot be defined holistically without the inclusion of ants and bees, an approach followed in this book, although both the ants and bees have been far more comprehensively treated in sister volumes to this series.

The name Hymenoptera means "membrane wings." This is not terribly informative, as many insects have membranous wings. However, the wasps, bees, and ants do possess a unique wing characteristic: hook-like projections called hamuli that couple the forewings and hindwings together in flight. One feature of the Hymenoptera that was key to their evolutionary success, and which is often very obvious, is the ovipositor, frequently acting as a sting.

The sawflies and woodwasps lack a wasp "waist" (there is no constriction between the anterior abdominal segments). All other Hymenoptera with a narrowed waist are wasps, but two groups of wasps have been recognized as morphologically and behaviorally distinct and given their own group names. Ants are always eusocial (*see also* page 19—some are socially parasitic, though) and recognizable by their wingless workers, a petiole of one or two node-like segments, and their elbowed antennae.

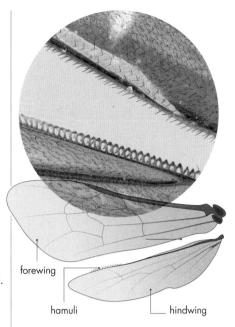

forewing

hamuli

hindwing

ABOVE | Close-up of the hook-like hamuli projecting from a short section of the leading edge of the hindwing that couples with the forewing in flight to create a single unit.

BELOW | A brilliantly colored cuckoo wasp (Chrysididae).

BELOW | **A.** A spider-hunting wasp (*Auplopus*). **B.** The California Gall Wasp (*Andricus quercuscalifornicus*), a gall-former on species of *Quercus* (the white oaks).

A.

B.

Species in many other families have evolved winglessness too but do not share all these characters. Bees are usually hairy (the hairs or setae visibly branched under the microscope), females (except in socially parasitic species) have pollen baskets which are specialized pollen-collecting structures on their hind legs or abdomen, and the mouthparts are modified into a nectar-sucking proboscis. Again, recognition of an insect as a bee is not always straightforward: various wasps can be similar, and some bees are mostly hairless. So that leaves the rest of the Hymenoptera as "wasps." Beware, some of these are wingless, and some are highly hairy.

ABOVE | Male (on top) and female of a seed predator parasitoid wasp (*Eurytoma*) about to mate before the female lays an egg into the seed capsule of a *Bulbine* species (Xanthorrhoeaceae). Her larvae will feed on the seeds, destroying them in the process.

ABOVE RIGHT | A Darwin parasitoid wasp (*Pimpla*). Species in this genus attack lepidopteran pupae, laying an egg directly into the chrysalis. The wasp larva develops inside the pupa.

PARASITOIDS

While the larger, more colorful wasps are often easily observed, visiting flowers or hunting, the majority of wasp species are tiny parasitoids and are overlooked or dismissed as midges or other insects. Parasitoids are wasps that use other insects (usually the immature stages) as a food resource for the development of their own offspring. The female parasitoid wasp either lays her egg or eggs into or onto the host insect, and her larvae will eventually kill the host by feeding on the body contents. The term parasitoid is used rather than parasite because there is a fundamental difference in the interaction with the host. Unlike parasitoids, parasites don't directly result in the death of the host through their feeding, although they may transmit pathogens that could kill the host.

Parasitoids are common inhabitants of all environments, including gardens and parks in suburbia and cities, where they can be amazingly diverse. Wasps are often doing us a service by controlling pest species. Appreciating these wasps is simply a matter of adjusting your mindset and observational focus to a smaller scale. Careful and closer examination of vegetation—such as under leaves, on top of the leaf litter beneath shrubs, or in among grasses and herbs in meadows or fields—will open a whole new world of fascinating species richness. Try it—it may astound you with a new appreciation for the little insects.

ANATOMY OF WASPS

The wasps, with the exception of the Symphyta (see page 16), deviate from the basic insect ground plan where the body is divided into the head, thorax, and abdomen. This is because the apparent division between the thorax and abdomen is after the first segment of the abdomen, which is fused to the thorax in the Apocrita (see page 61). The thorax is therefore named the mesosoma (thorax plus first abdominal segment) and the remaining abdomen termed the metasoma.

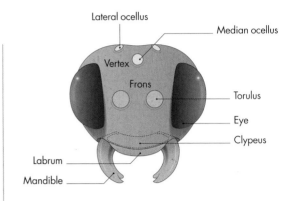

ABOVE | Typical wasp head in frontal view.

RIGHT | Typical symphytan wasp body in lateral view (petiole is absent).

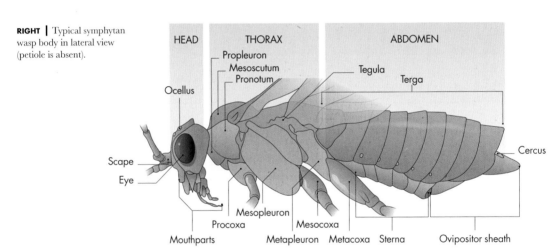

RIGHT | Typical apocritan wasp body in lateral view with main body parts labeled.

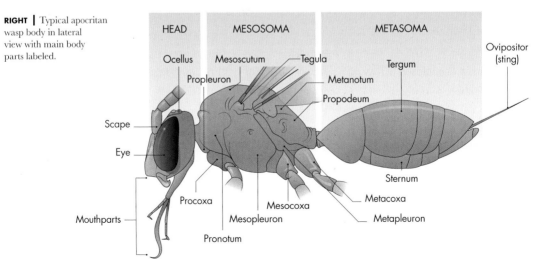

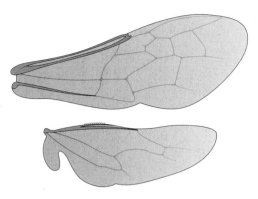

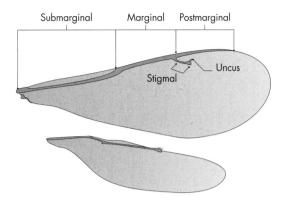

LEFT AND RIGHT | Generalized wing venation pattern of aculeate wasps (venation usually complete or nearly so), contrasting with that of micro (chalcid) parasitoid wasps, where the venation is often dramatically reduced.

WINGS

Wasps have two pairs of wings, but varying degrees of wing reduction or entire wing loss has evolved independently in many families. This is often an adaptation to their lifestyle. For example, several lineages of Scelionidae have lost their wings as they impede progress when burrowing through the silk of spider egg sacs to reach the host eggs for oviposition. Alternatively, it can be an adaptation to the habitat they live in. Male fig wasps, for example, do not need wings as they are adapted to spending their short lives of only a couple of hours inside the enclosed fig cavity, or parasitoid species inhabiting confined habitats such as leaf litter do not necessarily need to disperse long distances, and it is easier to negotiate such a substrate with a streamlined body. Wing venation is often reduced, extremely so in most parasitoid families, and may be related to the overall reduction in body size. Configuration of the veins is often important and diagnostic for separating families. The wing venation

pattern is the most convenient characteristic for separating the two most species-rich families, the Darwin wasps (Ichneumonidae) and their sister group, Braconidae.

HEAD

Wasps have chewing mouthparts, often with strong mandibles that may be bizarrely developed; extreme examples are those of males of the Sulawesi species *Megalara garuda* (Crabronidae), and some potter wasp (Vespidae) males in the genus *Synagris* have evolved tusk-like mandibles for battling other males to gain mating opportunity or to guard mud nests with developing larvae constructed by females. Males of the large (1⅜ in [3.5 cm]) African species *Synagris cornuta* may have mandibles as long as their forelegs. In some species, development of elongated mandibular tusks is combined with the development of a long clypeal horn, as in *Synagris fulva*. The impressive mandibles of some non-pollinating fig wasps are

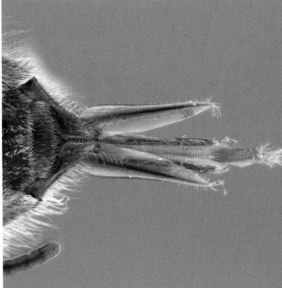

A.

B.

used to engage in gladiator-like lethal combat, often slicing their opponents in half. This evolution of aggressive secondary sexual characteristics is unusual for Hymenoptera. Many groups, including both wasps and bees, have evolved a long "proboscis" facilitating the drinking of nectar from flowers or the gathering of water for nest construction or evaporative cooling of the nest. The sensory mouthpart palps can be extremely elongate or highly reduced.

The two compound eyes are usually large, accompanied by three small simple eyes (ocelli), although both types have been reduced or lost in several groups—some fig wasp males and flat wasp males are effectively blind. Predatory species such as the Crabronidae and Sphecidae often have large compound eyes, sometimes meeting on top of their head, facilitating their navigation through complex landscapes at high speed in pursuit of suitable prey to provision their larvae. Nocturnal wasps typically have very large compound eyes and ocelli.

Antennae are used to "smell" and are variously modified. They have minute sensory protrusions and many pits, usually contained on plate-like or elongate sensilla to facilitate olfactory reception of volatile chemicals, necessary for host- or mate-finding. Some species, particularly in various males of Chalcidoidea, have complex branched antennae to increase their surface area for more efficient chemical volatile detection. Males of various groups, such as the Darwin wasp subfamily Diplazontinae, have thickened tyloids that enable the antennae to coil around the female's antennae.

LEGS

Legs may be modified as an adaptation to differing biology. The forelegs of sand-nesting species such as the digger wasps may have rakes of modified spines assisting with backward propulsion of sand during burrow excavation. Those of oil-collecting bees such as *Rediviva emdeorum* are exceptionally long, an adaptation for reaching down the twin spurs of *Diascia* flowers to reach the oil reward

C.

at the base of the corolla tube, in the process
pollinating the flower—an example of a
co-evolved pollination syndrome.

OVIPOSITOR

In the basal symphytans the ovipositor is saw-
like and is used to penetrate plant tissue for
egg-laying. The parasitoid wasps have a slender
ovipositor used for penetrating the cuticle of
their host insects and/or the substrate
surrounding the host for egg-laying. In most
cases, the venom is used for host immobilization
or modification of behavior, either temporarily
or permanently. The ovipositor can be extremely
long, particularly in those parasitoids that need to
reach host larvae feeding deep inside tree trunks,
and in some cases (*Zaglyptogastra*, Braconidae) can
be serially arched, an adaptation for enabling
negotiation down a convoluted host larval tunnel.
The social wasps have evolved potent venom that
is used to defend their nest resources from
predation by vertebrates or attacks by other
insects. In these stinging wasps and in parasitoids

that directly contact the host, the eggs are laid
from the base of the ovipositor, rather than from
the tip as in other parasitoid wasps.

LARVAE

The larvae of parasitoid and aculeate wasps are
typically maggot-like, without legs, developing
in sheltered situations, often inside their host or
attached to the outside of a host that is feeding
in a concealed place. Their gut remains sealed,
preventing contamination of their surrounding
environment, with the fecal contents only released
prior to pupation. Some parasitoids have planidial
first-instar larvae that need to be mobile to find their
host, because the female lays her eggs away from
the host. Sawflies feeding externally on plants have
caterpillar-like larvae with a defined head and large
mandibles, and thoracic legs with a set of prolegs
present on the underside of each abdominal
segment. They can be told apart from caterpillars
by the absence of crochet hooks on the prolegs.
Caterpillars have prolegs on fewer segments.

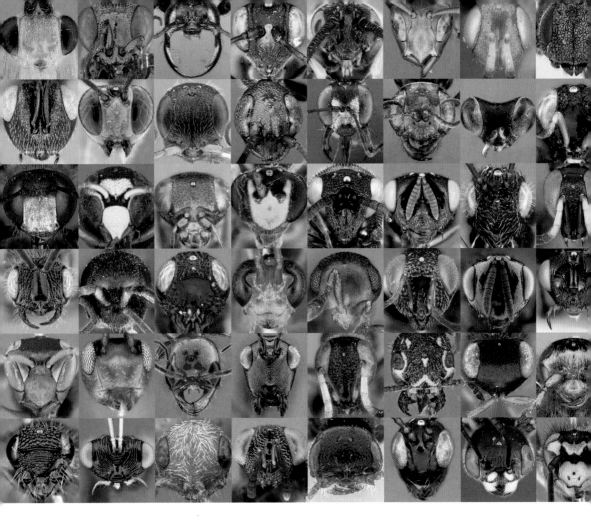

DIVERSITY OF WASPS

ABOVE | A diversity of
wasp head shape and color.

There are about 154,000 described species and 2,400 extinct
species (described from fossils) in the Hymenoptera, classified
in about 8,500 genera (about 690 extinct genera) currently
placed in 143 extant families, but this figure is dynamic and
often changes as new phylogenetic assessments of higher
wasp classification are undertaken. Most species are still
undiscovered, particularly in the tropics, and the true species
total is probably 10 to 20 times higher than these figures.

   With 143 different families of wasps, bees, and ants,
there is a great range of different physical forms and biology
present in this order of insects. There are currently 29
superfamilies, a higher classificatory category comprising

related families of wasps. The ancient plant-feeding sawflies and woodwasps, Symphyta, comprise eight of the superfamilies but by far the fewest species (about 8,100 described species) of the traditional subdivisions of Hymenoptera. Ichneumonoidea is the largest superfamily of Hymenoptera, containing about 45,000 described species in the two most species-rich wasp families: Braconidae and Ichneumonidae. In some of the better-known faunas, such as Europe and North America, around 20 percent of all insect species are ichneumonoid wasps, which play a hugely important role in terrestrial food webs, as parasitoids of many other insects.

Although often with small and difficult-to-identify species, the three next largest superfamilies, Chalcidoidea (about 23,000 species), Cynipoidea (about 3,200 species), and Platygastroidea (about 5,400 species), are not only economically important, also containing mostly parasitoid species, but are exceedingly interesting superfamilies from an evolutionary perspective, with major shifts in hosts between families. The jewel wasps (Chalcidoidea) are a huge superfamily of often tiny but colorful wasps. This is probably the wasp superfamily with the greatest variation in morphology and includes a number of independent reversals to plant-feeding from a parasitoid lifestyle, some of the better-known examples being the fig wasps (Agaonidae and Pteromalidae) and seed chalcids (Megastigmidae). Cynipoidea include one major radiation of species—the gall wasps—that have reversed from a parasitoid lifestyle to one of plant-feeding. Platygastroidea, as is true for the Chalcidoidea, attack a huge range of host insects and spiders, but most of the species in these superfamilies are still undiscovered, and for those species that are named their host relationships are mostly unknown or poorly understood.

The remaining less diverse parasitoid groups are of course equally interesting and important from evolutionary and biological perspectives. The aculeate wasps comprise mostly predatory species and contain the several groups of truly social species, the ants, some bees, and paper wasps, with about 67,000 described species. Given that the aculeate wasps are mostly larger species with conspicuous nesting, flower-visiting, and hunting behaviors, they are better known than the parasitoid groups, but still much remains to be discovered and documented in this group as well.

FUTURE DISCOVERIES

The task of discovering and documenting the unknown species of wasps entails expeditions to many remote, unexplored areas of the world, but also continues in our own backyards. A diverse array of collecting methods is used, including Malaise traps, sweeping, yellow-pan traps, pitfall traps, leaf-litter extraction, light trapping, rearing, and the like, each method targeting a different assemblage of species.

The resulting specimens are curated and accessioned into natural history museum collections where they are available for taxonomic research and biodiversity assessment. The majority of species are only millimeters in length, and this small size, combined with high diversity, means that new species are easily discovered.

Taxonomy, the science of describing species and their relationships, is a hugely important topic in entomology, given the vast numbers of species that still mostly lack names. This primary discovery is essential to unraveling the biology, evolution, and ecological relationships of these insects. The fascinating life histories we know about must only be scraping the surface of understanding the world we live in.

EVOLUTION OF WASPS

Wasps are ancient, having been around since a little before the earliest days of the dinosaurs. The Hymenoptera arose from a common ancestor shared with the Panorpida lineage that gave rise to the flies, fleas, caddisflies, butterflies, and moths and relatives. Wasps first appeared in the fossil record in the middle of the Triassic period (about 230 million years ago, or MYA), but DNA evidence suggests that Hymenoptera originated in the Permian. The oldest known fossils belong to the weird-antennae sawflies (Xyelidae), with the main wasp groups having evolved by the late Jurassic (155 MYA). The parasitoids and the stinging wasps likely radiated along with the major diversification of insects and flowering plants during the Cretaceous period, 145–65 MYA.

ABOVE | A wasp fossil, *Palaeovespa florissantia* (Vespidae), preserved in the 34-million-year-old (Eocene) Florissant formation comprising shale and mudstone beds in Colorado (USA).

OPPOSITE | A Darwin wasp (*Dolichomitus* species) with a 2³⁄₈ in (6 cm) long ovipositor used to reach the larvae of beetles deep in their galleries. Parasitoids with long ovipositors need to manipulate their metasoma in a sequential pattern of varying convolutions to optimally position the slender ovipositor for drilling or threading into the host substrate, often using their hind legs to help guide the process.

EARLY WASPS

The first wasps were all plant-feeding in their larval stages, with the evolution of the parasitoid (210 MYA) and stinging/predatory (155 MYA) modes of life only evolving later. The group of sawflies, woodwasps, and horntails (the symphytans) were the first wasps that arose, and many of their lineages are only represented in the fossil record, suggesting that they used to be far more diverse. Nevertheless, many families still successfully exist today, and groups such as the Xyelidae could be termed "living fossils" in much the same way as cycads, ginkgos, coelacanths, lungfishes, and the Platypus (*Ornithorhynchus anatinus*) are. The parasitoid woodwasps (Orussidae) are the only symphytans that have a parasitoid lifestyle, and this lineage is sister to the narrow-waisted wasp (apocritan) group, which comprises most of the extant wasp species richness (about 95 percent of all wasps).

OVIPOSITOR EVOLUTION

Major features that have led to the immense diversification of Hymenoptera include the retention of an external ovipositor with interlocking sections and a haplodiploid sex-determination system. The ovipositor has aided the exploitation of all sorts of food resources; it has evolved from laying eggs in plant tissue to becoming a multi-purpose tool, drilling, injecting venom, and laying eggs. The ovipositor and some biochemically complex secretions associated with oviposition were probably key to the diversification of wasps, as the common ancestor of all the non-symphytans was a parasitoid of other insects, accessing concealed hosts via the ovipositor and stinging them. In some parasitoids the ovipositor has evolved to be a means of precisely depositing an egg in particular host tissues. In the stinging aculeates and some other parasitoid groups that lay eggs on exposed hosts, the egg no longer passes through the ovipositor but instead issues from the genital opening at the base of the "ovipositor," which then acts only as a sting. Venoms clearly evolved to manipulate hosts by paralyzing them to keep them fresh as food for offspring but have diversified in chemistry and function. The use of a sting to deliver venom is one of the reasons for the success of hunting wasps and the evolution of eusociality, stings being critical to the defense of nests and colonies by wasps, bees, and ants.

PARASITOID ADAPTATION

Parasitoids need to overcome the immune response of the host to their egg that has been laid inside the host's body. This is usually achieved through the simultaneous injection of proteins during egg-laying, but some Braconidae and Darwin wasps have associated polydnaviruses that perform this task of circumventing the immune response and that also alter the host's development and metabolism, benefiting the development of the parasitoid larva. Bracoviruses and ichnoviruses have separate ancient (around 73-million-year-old) symbiotic associations with Braconidae and Ichneumonidae, possibly explaining the huge and

successful species radiation of these parasitoid wasps. Many of the parasitoid wasp families also harbor *Wolbachia*, intracellular parasitic bacteria inherited through the maternal line that skew the sex ratio of the wasp's offspring. Some wasps are not able to reproduce successfully without infection by the bacteria, and here the association is mutually beneficial rather than parasitic.

FURTHER EVOLUTIONARY DIVERSIFICATION

The majority of parasitoid wasps are solitary but there are many gregarious species, taking advantage of larger hosts to lay clutches of eggs. An extreme version of gregariousness has evolved at least twice in wasps that are polyembryonic. In these cases, best studied in the tiny jewel wasp *Copidosoma floridanum*, a single egg is laid that then cleaves into multiple embryos, giving rise to a clonal army of parasitoid wasps within the caterpillar host. Most aculeate wasps and bees are also solitary. Solitary species provide their offspring with food provisions and sometimes

a nest but never have any direct contact with their young. Ants and several groups of bees have independently evolved social behavior similar to the social paper wasps. Although ants and bees are readily distinguished by the layperson, both groups are in fact derived wasps closely related to other wasp families. True eusocial behavior, with a division of labor between reproductive and non-reproductive castes, has evolved independently at least 11 times in the order Hymenoptera, whereas in all other insects it has evolved only once (in the termites, which are basically eusocial cockroaches). Many social Hymenoptera (ants, bees, and wasps) have potent venom, which they deliver via a stinging apparatus at the tip of their abdomen to defend their nest against predators. Venoms were already used by the parasitoid wasps to paralyze hosts, with this venom co-opted as a defensive measure. In many groups the venom fulfills both purposes.

WASP BIOLOGY

The biology and life history strategies of the order Hymenoptera are incredibly varied, from the ancestral plant-feeding mode through to various parasitoid and predation strategies, with multiple reversals to plant-feeding, including numerous evolutions of gall-forming, and, rather famously, advanced eusociality to a degree only the termites can match. The majority of families contain parasitoids that attack other insects, some of which are pests of our agricultural and forestry industries. Through this process these wasps perform critical control of insect populations and pest species. Plant-feeding species are used in the biological control of invasive plants, although several sawflies have become notorious pests themselves. Several species of wasps are considered to be invasive pests or detrimental to biocontrol programs.

FOOD

Nutrition intake for development takes place exclusively during the larval stage, with this phase usually comprising the major portion of the wasp's lifespan. The sawflies are usually phytophagous, feeding exposed, like caterpillars, on leaves, flowers, pollen, or plant stems; parasitoid species feed on the body contents of their host; predatory wasps sting and provide their larvae with paralyzed prey; ants are usually predatory but also feed on cultivated fungi; and the bees feed on nectar and pollen. Adult wasps do not feed except to obtain input for their energy requirements and for maturation of their eggs (oogenesis). They will take in sugar-rich sources such as nectar, as well as hemolymph and other mineral-rich bodily fluids from their host insects—often imbibing fluids exuding from oviposition puncture wounds or through deliberate mastication of prey items—a prerequisite for egg development in those wasps that lay large, yolk-rich eggs.

BELOW | *Cotesia glomerata* (Braconidae) larvae recently emerged from their host Large Cabbage White (*Pieris brassicae*) caterpillar to spin a cocoon, wherein they pupate before emerging as an adult wasp. The parasitoid wasp larvae have devoured the haemolymph of the host caterpillar, dooming it to die, and hence are often used as effective biocontrol agents of pest insects.

Larvae of parasitoids usually end up consuming most of their host tissues, for example often leaving just the hard larval head capsule or pupal case, but some feed only on hemolymph, such as the very successful braconid subfamily Microgastrinae. Unlike the larvae, adult digestive tracts are not developed for dealing with intake of solid food sources. In contrast, numerous sawflies, such as some Tenthredinidae, are predatory as adults; these families do not have the constriction of the digestive tract that parasitoids and their descendants have. For the parasitoid wasps and their descendants, a protein-rich diet of fresh meat is essential. Pollen has become the protein source for bees and pollen wasps.

WASPS AS PREDATORS

Some aculeate wasps made the transition from a parasitoid to a predatory life history, provisioning their larvae within nests. This was a prerequisite for feeding larvae on pollen and nectar, and also a prerequisite for sociality, which evolved independently many times. Sociality within Hymenoptera has been categorized into five different levels (communal, quasisocial, semisocial, subsocial, and eusocial), representing increasing levels of cooperation between individuals within the colony. The better-known eusocial groups—ants, honey bees, bumblebees, stingless bees, hornets, paper wasps, and yellowjackets—form social colonies usually comprising a queen and many female workers that do not reproduce but spend their lives defending the nest and looking after and raising their sisters. Reproductive males and females are produced at particular times or under certain conditions to start new colonies. Males die soon after mating, and in the case of honey bees, immediately after mating as their genital capsule is explosively severed, to increase the chance of a male's sperm fertilizing eggs. The mated reproductive female will become queen of a new colony. She stores sperm for fertilization of her eggs, over several years for honey bees and many ant species.

HAPLODIPLOIDY

Haplodiploidy in Hymenoptera means that males are haploid (one set of chromosomes) and females diploid (two sets of chromosomes, one from each parent), and therefore males develop from unfertilized eggs. The consequence of this is that resources can be allocated differently depending on whether a female or male egg will be laid. In parasitoid wasps, this gives more flexibility in use of hosts and the ability to lay more female eggs than male, thus ensuring that scarce or sparse resources are used more efficiently. In social nesting aculeates, haplodiploidy has significant consequences for relatedness, with a female being more closely related to (sharing more genes with) her sisters than to her own offspring, thus setting some of the conditions for eusociality, forfeiting

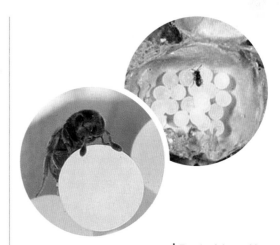

ABOVE | Female of the maritime parasitoid wasp *Echthrodesis lamorali* (Scelionidae) ovipositing into an egg of the host intertidal spider (*Desis formidabilis*).

RIGHT | Close-up of the brilliant refractive coloration of the Splendid Cuckoo Wasp (*Chrysis splendens*, Chrysididae).

the opportunity to mate and reproduce and instead raising more sisters as workers. The genetics are usually more complex, with queens of many eusocial species in fact mating with multiple males, meaning that the strength of relatedness between sisters is diluted, but most of the sperm is contributed by few males. Males are irrelevant for much of the time in social Hymenoptera and are produced only when mating is needed. Lacking a sting, they play no role in colony defense.

COLORFUL WASPS

Why are some wasps so colorful? Bright contrasting colors, often combinations of black, red, white, or orange, are used as a visual signal, an aposematic warning to potential predators of their painful sting. Velvet ants, which have a potent venom, often have patterns with these colors and the flightless females run around freely during the day, exposed to predators. Warning coloration is sometimes combined with sound production,

such as the squeaks of many velvet ants, or in some of the metallic blue spider-hunting wasps in the genus *Hemipepsis*, which produce a clacking sound in flight. Bright color can also be related to their biology. Iridescent coloration, such as in cuckoo wasps (Chrysididae), is a result of light refraction.

The range of refracted color is determined by the structure of the extremely hard exoskeleton, which is constructed as a stack of thin chitin layers separated by very thin irregular air gaps or by a helicoidal arrangement of chitin fibrils. These complex structures break up the white light spectrum creating a variety of refractive indices. Cuckoo wasps enter other solitary bee and wasp nests to lay their own egg, and they need a strong exoskeleton as a defense against the host sting if they are discovered. Risky behavior, but it saves the female having to provide for her own young, as her larva consumes both the host larva and the host's food provision.

UNDERWATER WASPS

Most wasps are terrestrial, but there are a few (around 150 species) that are aquatic or semi-aquatic. The scelionid *Echthrodesis lamorali* is an egg parasitoid of intertidal spiders that live submerged in the sea inside retreats made in old shells, commonly limpets, trapped under boulders in the intertidal zone. They spend most of their lives completely submerged in this turbulent environment, surviving in the air trapped within the retreats, only becoming exposed to the air during spring low tides. The wasp has a coating of dense hairs, speculated to trap air in a surrounding bubble, and has lost its wings as they are unnecessary in its specialized habitat, and this streamlining of the body also helps with burrowing into and out of the silken spider egg sac.

The diapriid *Trichopria columbiana* is a parasitoid of aquatic *Hydrellia* flies (Ephydridae). The female wasp traps an air bubble under her wings allowing her to continue breathing through her spiracles while swimming down to locate the host fly puparia for egg-laying. *Agriotypus* species are Darwin wasps that enter the water and seek caddis pupae to parasitize, gripping vegetation with long claws and, again, trapping air in a plastron around the body, thanks to dense setae. The larva of *Agriotypus* spins a long silken tube that is expelled from the host caddis case, acting as a gas exchange mechanism while the wasp pupates.

PARASITOID LIFE

The parasitoid mode of life usually relies on concealing the developing wasp larvae, either inside the body of the host or within the retreat or substrate that the host is developing in, with the parasitoid larvae developing by feeding externally on the host, reducing the need to overcome the host's immune responses. However, several groups of parasitoid wasps have reversed to a concealed plant-feeding mode, developing inside seeds or inside a modified plant part, termed a gall.

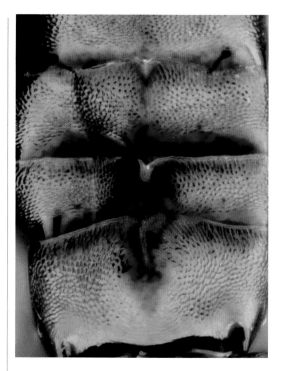

A gall is an abnormal growth formed by a plant in response to the attack of other life-forms, and can comprise a great diversity of shapes, colors, and sizes, the best-known wasp galls probably being the cynipid oak galls. Although gall-formers feed in concealment, this is not a safe existence as galls are attacked by large numbers of parasitoids; indeed, cynipid gall-formers, and probably other gall-formers, may have evolved from parasitoids of other insect galls.

Another example of parasitoids changing back to plant-feeding is the relationship between fig wasps and their host figs, a classic example of an obligate mutualism with the fig wasps developing in galled florets inside the fig, with these tiny wasps in turn providing a specialized pollination service for the fig trees. This interaction is then exploited by parasitoid wasps that have evolved extremely long ovipositors allowing the female parasitoid to access the host fig wasp larvae developing deep within the fig from the outside of the fig wall.

LEFT | Color detail of the metasoma of a parasitoid fig wasp (*Sycoscapter*, Pteromalidae), a result of structural light refraction rather than color pigment.

BELOW LEFT TOP | Pollinator fig wasp females (*Courtella*, Agaonidae) forcing their way through the ostiole of *Ficus modesta* to reach the florets lining the central fig cavity to perform pollination and oviposition.

BELOW LEFT BOTTOM | Galls formed on the underside of an oak leaf by gall wasps (Cynipidae), providing a protected resource for development of their larvae.

RIGHT | The North American Black Giant Darwin wasp (*Megarhyssa atrata*), a parasitoid of the wood-boring larvae of the woodwasp *Tremex columba* that feed inside dead tree trunks.

Many parasitoids in other wasp families have also evolved long ovipositors. The most spectacular species are *Megarhyssa* Darwin wasps, parasitoids of deeply concealed woodwasp larvae and pupae. Locating a host deep within solid wood comes with challenges. The host is not directly located; instead, the *Megarhyssa* detects a fungus with which the ovipositing woodwasp inoculates the wood. The fungus breaks down cellulose (which very few insects can digest) and is eaten by the woodwasp grub. The female *Megarhyssa* needs to manipulate its very long and rather flexible ovipositor to be able to drill through the wood. It is braced along the underside of the metasoma, and in many wasps this is achieved by standing tall on the long hind legs.

However, *Megarhyssa* have such long ovipositors that they can only brace the drilling section by rotating the apical metasomal segments by nearly 360 degrees, pushing the ovipositor up into a membranous sac that protrudes as the ovipositor coils up. With the exposed section of the ovipositor thus shortened, drilling can commence. It can take a couple of hours to drill down to the host and often they miss, necessitating the withdrawal of the ovipositor and drilling again. When they do contact a host, it is stung, and permanently paralyzed, with the egg squeezed down the ovipositor and laid on or next to the host. The size of the host can be estimated by the female just by the contact with the ovipositor; larger insects will host females, while smaller insects will host males. Males do not need to be so big, so good resources do not need to be wasted on them.

These reflections depict only a few examples of the immensely varied biology of Hymenoptera. Further lifestyle strategies that have evolved are highlighted under the respective family treatments through the book.

CONSERVATION OF WASPS

Wasps, bees, and ants are fundamental to the continued existence of healthy ecosystems. They provide essential ecosystem services, via pollination, predation, and recycling of nutrients and carbon. Without parasitoid and predatory wasps and their diet of other insects, we would have a much more imbalanced world. Parasitoid wasps play a vital ecological role as natural controllers of insect populations, including species that are detrimental to agricultural and forestry industries, and have vast potential for use in managed biocontrol programs. Many wasps are important pollinators, alongside bees, but their contribution to pollination has been insufficiently explored, which is also true of flies, beetles, moths,

and various other under-appreciated groups of insects. There is no doubt that they play a critical role in the maintenance and evolution of floral species richness. The better-known groups have potential as indicator species in conservation and ecological monitoring.

INVASIVE WASPS

Even though most wasps are beneficial, a few species have become pests outside of their native range. The invasive German Yellowjacket (*Vespula germanica*) has been inadvertently introduced to many parts of the world including Argentina, Australia, Chile, New Zealand, South Africa, remote Saint Helena, and the USA. This is

LEFT | European Paper Wasp (*Polistes dominula*), a Palearctic and Oriental species, introduced to Africa, Australia, and the Americas. It is a serious invasive species competing directly with indigenous paper wasp species, disrupting natural ecosystem processes. It is social and constructs exposed communal paper nests, which can reach a much larger size than indigenous paper wasp nests.

a eusocial species, constructing concealed paper nests that can reach a massive size, often with many thousands of individuals when the climate is mild. This species is a pest and incidental nuisance to picnickers as workers feed on sugary substances and have a painful sting that is used readily if disturbed.

More serious impacts are focused on local ecosystems through competition and preying on large numbers of native insects; on the honey industry as they hunt bees; and on the wine and fruit industry through fruit-feeding damage and subsequent secondary fungal infection. The European Paper Wasp (*Polistes dominula*) is a social wasp that has been widely introduced to Africa, Australia, and the Americas, competing directly with indigenous paper wasp species and in the process disrupting natural ecosystem processes.

The Asian Northern Giant Hornet (*Vespa mandarinia*, unfortunately nicknamed "Murder Hornet") is currently creating concern in North America, where it has potential to impact honey bee colonies. The Asian Yellow-legged Hornet (*Vespa velutina*) is already established over much of southern and central Europe and is another specialist honey bee predator. It should be noted, though, that these eusocial vespine wasps are also model organisms for studying the evolution of complex behaviors, including building their amazing nests.

Although huge numbers of wasp species are specialized in attacking other insects and are thus near the tops of their food chains, operating at high trophic levels, we have very little data on

ABOVE | A male *Podalonia canescens* pollinating the sexually deceptive orchid *Disa atricapilla*. There are multiple pollinia attached to the wasp's underside.

populations of any wasps. Given the levels of concern over general insect declines, the fate of wasps, including differences between host specialists and generalists, is a field ripe for research.

Proactive actions, such as erecting bee hotels in gardens and parks, leaving weeds in place that are actually indigenous wildflowers, and landscaping with indigenous plants, will all help to maintain insect populations, and ultimately help to mitigate the decline in our insect faunas. Ultimately, the populations of wasps depend on those of other insects, and all depend on sufficient, connected habitat, which requires concerted effort in a rapidly changing world.

WASP PHOTOGRAPHY

Most wasp species are tiny. As a fully formed adult wasp, many are only $^1/_{32}$–$^1/_{16}$ in (1–2 mm) in length, necessitating expensive macro equipment or microscope facilities to photograph these species. Photographic documentation of these minuscule insects has associated technical difficulties. The high magnification required to observe and photograph small objects reduces the depth of field (or thickness of the plane of focus), with the result that a single photograph may capture part of the insect in sharp focus, but the rest of the insect will be blurry.

STACKING SYSTEMS FOR FOCUS

Various types of automated or manual multi-stacking imaging systems have been developed over the last 20 or so years. Systems such as these are increasingly being used to produce close-up images of museum specimens. These stacking systems use a process where several images are taken at different focal planes through the depth of the insect. The operator sets the top focal plane (the closest part of the insect to the microscope lens) as well as the bottom focal plane (the furthest part of the insect away from the lens), and an automated Z-stepper (this can also be done manually) moves the lens in even increments between these two settings. A camera can be used attached to a microscope or alone, with a macro-lens or microscope objectives. The resulting images are combined into a single image by a computer algorithm, resulting in a fully focused high-magnification, sharply defined image of the specimen.

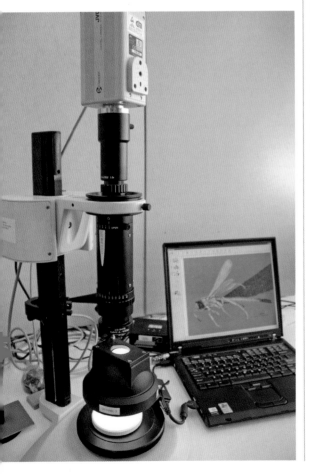

LEFT AND ABOVE RIGHT | One example of a multi-stacking imaging system (left) with an example (right) of a resulting series of images, captured at different focal planes down through the specimen, prior to amalgamation into a single, fully focused image.

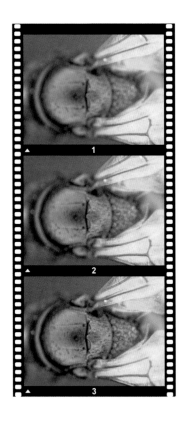
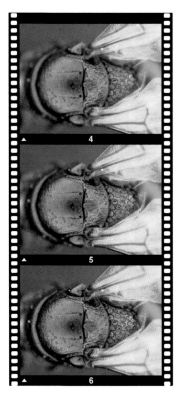
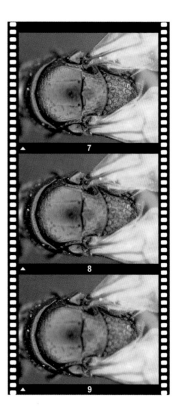

PHOTOGRAPHY IN THE FIELD

This works fine for static preserved specimens but photographing wasps in the wild takes extra patience and dedication using handheld macro SLR camera setups with flash systems operating at exposures anywhere between 1/60th and 1/200th of a second to freeze the subject's movement. Using a bright flash rather than daylight also allows for greater depth of field, because of the higher light intensity. Alternatively, living specimens can also be photographed under controlled environmental conditions in light boxes and at cold temperatures to reduce their movement.

With the rapid technological advancement in smartphones, many models now have high-quality macro capabilities facilitating easier photography of insects, making this capacity readily available to the layperson. This has the added advantage of enabling rapid uploading of captured images to citizen science platforms such as iNaturalist, in the process assisting with documentation of species and their geographical distribution, which is a valuable contribution to the development of biodiversity knowledge. There also exists the opportunity to capture fascinating aspects of wasp biology, a novel ground-breaking pursuit given that the biology of most species is unknown or poorly recorded. Wasp photography can be incredibly creative and rewarding.

CLASSIFICATION OF WASPS

With the advent of additional molecular tools, the classification of Hymenoptera is rapidly undergoing revision, and various families and superfamilies are being newly defined or redefined. Increasing analysis based on increasingly large amounts of genetic data is providing insight into the complex patterns of morphological evolution, with previously shared character systems being shown to comprise more and more examples of convergent evolution as a response to similar underlying evolutionary drivers based on ecological or environmental adaptation. This is particularly true of the aculeate wasps and the jewel wasps.

BELOW | A simplified phylogenetic tree depicting one interpretation of the evolutionary relationships of the hymenopteran superfamilies, illustrating that the Symphyta and Parasitica are artificially defined groups, each not defined by a common ancestor.

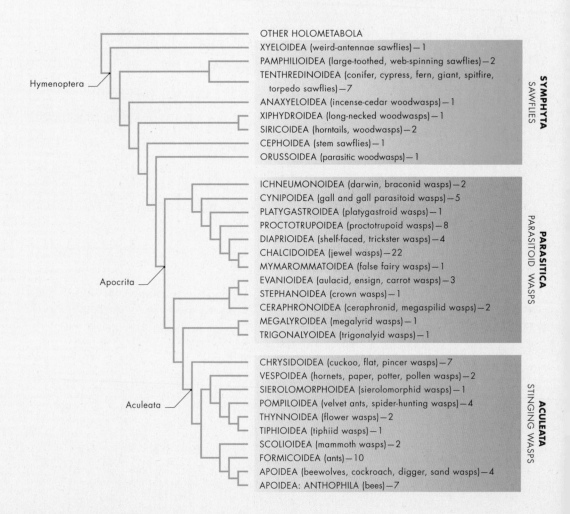

OTHER HOLOMETABOLA
XYELOIDEA (weird-antennae sawflies)—1
PAMPHILIOIDEA (large-toothed, web-spinning sawflies)—2
TENTHREDINOIDEA (conifer, cypress, fern, giant, spitfire, torpedo sawflies)—7
ANAXYELOIDEA (incense-cedar woodwasps)—1
XIPHYDROIDEA (long-necked woodwasps)—1
SIRICOIDEA (horntails, woodwasps)—2
CEPHOIDEA (stem sawflies)—1
ORUSSOIDEA (parasitic woodwasps)—1

SYMPHYTA
SAWFLIES

ICHNEUMONOIDEA (darwin, braconid wasps)—2
CYNIPOIDEA (gall and gall parasitoid wasps)—5
PLATYGASTROIDEA (platygastroid wasps)—1
PROCTOTRUPOIDEA (proctotrupoid wasps)—8
DIAPRIOIDEA (shelf-faced, trickster wasps)—4
CHALCIDOIDEA (jewel wasps)—22
MYMAROMMATOIDEA (false fairy wasps)—1
EVANIOIDEA (aulacid, ensign, carrot wasps)—3
STEPHANOIDEA (crown wasps)—1
CERAPHRONOIDEA (ceraphronid, megaspilid wasps)—2
MEGALYROIDEA (megalyrid wasps)—1
TRIGONALYOIDEA (trigonalyid wasps)—1

PARASITICA
PARASITOID WASPS

CHRYSIDOIDEA (cuckoo, flat, pincer wasps)—7
VESPOIDEA (hornets, paper, potter, pollen wasps)—2
SIEROLOMORPHOIDEA (sierolomorphid wasps)—1
POMPILOIDEA (velvet ants, spider-hunting wasps)—4
THYNNOIDEA (flower wasps)—2
TIPHIOIDEA (tiphiid wasps)—1
SCOLIOIDEA (mammoth wasps)—2
FORMICOIDEA (ants)—10
APOIDEA (beewolves, cockroach, digger, sand wasps)—4
APOIDEA: ANTHOPHILA (bees)—7

ACULEATA
STINGING WASPS

Hymenoptera

Apocrita

Aculeata

The higher classification of the jewel wasps (Chalcidoidea) has recently been revised based on molecular and morphological reassessment of evolutionary relationships, and 26 new families were circumscribed while this book was in production. We have comprehensively treated the jewel wasp families recognized at the time of writing and have indicated under the respective families from where the new families emanated. We also list these new families and their constituent genera on pages 152–54, but were unable to treat them to the same level as has been done for the other families.

The higher classification of the Aculeata has also been undergoing major reassessment based on molecular phylogenetic analyses. We have followed the latest proposed classification based on these findings. Not all authorities accept these proposed changes, as in some cases they remain to be corroborated by morphological phylogenetic reassessment. While the morphology of jewel wasps has undergone extreme diversification, the morphology of the other huge superfamily of parasitoid wasps, the Ichneumonoidea, is much more conserved and, barring occasional attempts to split off some groups, they remain classified in just two very large families.

LEFT | A pincer wasp female (Dryinidae).

DIRECTORY OF WASP FAMILIES

The directory section of this book that follows treats the wasp superfamilies in alphabetical order within each of the three higher categories, which are ordered according to their evolutionary age: sawflies and woodwasps (Symphyta); parasitoid wasps (Parasitica); and stinging predatory and provisioning wasps, bees, and ants (Aculeata). Within each superfamily the constituent families are arranged alphabetically, and therefore the superfamily and family arrangement does not reflect their evolutionary relationship to one another.

Larger families are treated in more detail than smaller families. Several families are rare, and their biology and distribution often remain to be elucidated, but all wasp families are included in this book. An overview is provided for each family highlighting notable behavior, biology, and typical characteristics, and an information panel summarizing distribution, constituent genera, habitats, size, activity, reproduction, and diet is included for each family.

SYMPHYTA
SAWFLIES AND WOODWASPS

The sawflies and woodwasps comprise several distinct clades of primarily phytophagous insects, although some of the woodwasps are effectively largely mycophagous. These groups are often collectively referred to as the suborder Symphyta, but with the other Hymenoptera clearly derived from within the Symphyta, this would necessitate the simultaneous recognition of monophyletic and non-monophyletic suborders, so we do not use the term suborders here (see the phylogenetic tree on page 30), but have retained the artificial historical grouping at this level for convenience sake.

Sawflies and woodwasps have ovipositors with conspicuous teeth, or annuli. These are used in most families for making a slit in plant tissue in which to deposit the egg, but in the woodwasps the ovipositor is used as a drill to penetrate solid wood. As with all Hymenoptera (except those with reduced wings), the symphytans have hamuli on the hindwing that lock onto the rear edge of the forewing in flight. Unlike the remaining Hymenoptera (the Apocrita), there is no sharp constriction between abdominal segments (although there is a slight constriction in Cephidae), so there is no "wasp waist." The tarsi of symphytans have small projections at the base of each segment, called plantulae, which are absent from most other Hymenoptera, although present in Trigonalyoidea.

Sawflies and woodwasps have a long history; the origins of Hymenoptera are thought to go back to the Permian, along with the other mega-diverse orders of holometabolous insects. The earliest branching Hymenoptera are the Xyelidae, with larvae that eat conifer seeds and with a fossil record stretching back to the Triassic.

LEFT | The sawfly *Cephus spinipes* (Cephidae), a stem-boring sawfly, drinking rain water from Red Clover (*Trifolium pratense*).

The largest radiations of symphytans have been associated with angiosperms, although a few species eat ferns, including the small family Blasticotomidae, and a pergid sawfly is known to eat fungi. Woodwasps develop in timber of conifers and angiosperms, consuming mainly saprophytic fungi—at least in species that have been well studied. Larval sawflies often feed externally on leaves, others feed inside plant tissue, in some cases inducing galls, or mining leaves; in many ways, the larval feeding habits of sawflies and Lepidoptera have followed similar pathways. The diets of sawflies and woodwasps have sometimes brought them into conflict with human agriculture and arboriculture, particularly when they have been inadvertently introduced outside their native range. As part of the effort to develop biological control of pest sawfly and woodwasp species, their natural enemy parasitoids have often been intensively studied, which has given us an invaluable insight into life histories of some parasitoid wasps, particularly the family Ichneumonidae.

Adult sawflies are frequently carnivorous, especially in the family Tenthredinidae, although some sawflies, such as some pergids, are very short-lived as adults and do not feed in that time. The Orussidae, a relatively small family of symphytan morphology, are parasitoids, and apparently the sister group to all other non-symphytan Hymenoptera. The orussids lack the wasp waist and, where known, kill the host rather than paralyze it, thus illustrating some potential traits of the ancestral parasitoid wasps.

LEFT | European Rose Sawfly
(*Arge ochropus*, Argidae).

INCENSE-CEDAR WOODWASPS

This family represents an ancient evolutionary lineage, with most genera (18) and species only known from the Mesozoic fossil record.

Syntexis libocedrii is the only living representative of the ancient family Anaxyelidae, and also the superfamily Anaxyeloidea, and is placed in the subfamily Syntexinae, which is best represented by eight genera only known from the fossil record.

The nominate subfamily Anaxyelinae is only represented by seven extinct genera, the Dolichostigmatinae by one and the Kempendajinae by two. Fossil species are known from as far back as the Middle Jurassic (174.1–163.5 MYA), but the peak of species richness is best represented in the Early Cretaceous (145–100.5 MYA).

The living species are restricted to the western part of North America from central California to southern British Columbia. These woodwasps are rarely seen, with specimens only usually being noticed soon after forest fires, when the adult wasps are most active.

The larvae are xylophagous (wood-feeding), developing in trees of *Calocedrus* (Incense-Cedar), *Juniperus* (Juniper), or *Thuja* (Cedar). Females only lay eggs into recently burned trees that can still be smoldering when the female inserts her ovipositor; eggs are laid in the wood just below the bark and small trees are preferred. Larvae develop inside the tree trunks, taking one to three years to complete development.

LEFT | Female *Syntexis libocedrii*, the only living representative of a previously diverse family of woodwasps, and a beneficiary of limited forest fires.

RIGHT | Female Meadowsweet Sawfly (*Phylloecus xanthostoma*), a fairly common and widespread European stem sawfly that mimics stinging wasps.

DISTRIBUTION
Only known from the mountains of western North America

GENUS
Syntexis

HABITATS
Forested mountains of western North America

SIZE
5/16–5/8 in (8–16 mm)

ACTIVITY
Diurnal

REPRODUCTION
Mating is brief, apparently without courtship

DIET
Larvae feed on wood of recently burned cedar or juniper trees, being best known for attacking Incense-Cedar (*Calocedrus decurrens*)

STEM SAWFLIES

The family Cephidae is most species-rich in the temperate northern hemisphere, with only a few species found in tropical or subtropical areas, namely the two species of the endemic Madagascan genus *Athetocephus*, the one species of the endemic Australian genus *Australcephus*, the one species of the endemic Indonesian genus *Sulawesius* and an Indonesian species of *Janus*. *Hartigia* species extend south just into the Neotropics (Mexico and Guatemala). There are about 23 genera and 160 species, arranged in 3 subfamilies: Athetocephinae, Australcephinae, and Cephinae, the latter containing most of the species. The extinct family Sepulcidae, with 15 genera, is the only other family assigned to the superfamily Cephoidea.

Unlike most other sawflies, these are slender insects. Prior to Orussidae being established as the sister group to parasitoid wasps, cephids were sometimes assumed to be the nearest relatives, because of their hints of a wasp waist and the absence of cenchri, a characteristic sawfly feature that is also absent in the Apocrita. The plant-feeding larvae bore inside grass stems or twigs of woody plants. The Wheat Stem Sawfly (*Cephus cinctus*) is a major pest of cultivated cereals in North America, following its introduction from Europe, and *Stenocephus fraxini* is considered a pest of ornamental street-shading trees in China. In both cases, these are examples of sawflies adopting introduced (and economically valuable) plants into their diet. Another major pest of wheat in North America is *Cephus pygmeus*, introduced from Europe.

DISTRIBUTION
Centered in the northern hemisphere; a few species in the southern hemisphere

GENERA
23 genera

HABITATS
Wide range, from grassland to forest

SIZE
Average $3/8$ in (10 mm) body length, up to $3/4$ in (20 mm) in a few species

ACTIVITY
Diurnal. Species feeding on grasses can often be swept in good numbers

REPRODUCTION
One generation per year. Eggs are laid directly into the host plant tissue. Pupation within the plant tissue, adults emerge in spring

DIET
Phytophagous. Larvae feed on tissue in grass stems, or shrub and tree twigs

PARASITIC WOODWASPS

The family Orussidae is most species-rich in the southern hemisphere. There are 16 extant genera and 82 species, with 1 fossil genus. They tend to be rarely collected.

Orussids are unique within the Symphyta, being the only parasitoid family within these otherwise plant-feeding Hymenoptera. They represent a sister group to the hyper-diverse radiation of apocritan wasps, with their wasp waists. Orussids represent a physiologically primitive branch of parasitoid wasps, perhaps close to the original parasitoid life history.

However, they are very specialized morphologically to their niche of parasitizing hosts deeply concealed in wood. The larvae attack xylophagous (wood-feeding) beetle or wasp larvae, commonly of jewel beetles (Buprestidae), long-horned beetles (Cerambycidae), and woodwasps (Siricidae, Xiphydriidae) developing inside dead tree trunks. Females locate the host larva feeding deep inside wood by tapping their antennae on the surface to generate vibrations that then bounce off the host, being detected by swollen subgenual organs in the forelegs—essentially echolocation, but via a solid medium. The ovipositor is long and coiled internally, being extruded a very short distance to allow drilling without the ovipositor buckling. The wasp larva initially develops externally on the host larva but later enters the dead host body to complete feeding.

The adult head has a characteristic crown of teeth circling the middle ocellus (simple eye). Many species have conspicuous color patterns, including red markings and patterned wings, and some are brilliantly metallic, as in *Chalinus braunsi*.

LEFT | Female of a spectacular African parasitoid woodwasp, *Chalinus plumicornis*, found widely in East Africa and the only *Chalinus* known in its range.

DISTRIBUTION
Worldwide, but highest diversity is centered in the southern hemisphere

GENERA
16 genera, including *Chalinus*, *Guiglia*, *Ophrynopus*, and *Orussus*

HABITATS
Wide range, from arid semi-desert habitats to rain forest

SIZE
$1/16$–$7/8$ in (2–23 mm) body length

ACTIVITY
Diurnal. Active in hottest part of the day

REPRODUCTION
Eggs are laid directly onto the host larvae using a very long ovipositor that is drilled into the tree trunk

DIET
Larvae feed on host larvae feeding inside dead tree branches

LARGE-TOOTHED SAWFLIES

BELOW | *Megalodontes* species on a flower. Despite being rather large and spectacular sawflies, *Megalodontes* species are difficult to identify.

The family Megalodontesidae is confined to the Palearctic region, frequenting temperate areas of Asia and Europe. It comprises a single genus (*Megalodontes*) with 42 species.

As in the Pamphiliidae these are wide-bodied, flattened wasps, but their antennae are much shorter. They are often brightly colored in combinations of black with yellow or white stripes and probably mimic stinging wasps for protection from predation.

The conspicuous adults fly strongly. Both sexes often cluster together at night in sleeping roosts. The large mandibles might be used for defense against other insects. Mating occurs on flowers and each mating can last several hours, mostly at dusk or dawn. Eggs are "glued" to the plant surface or laid in an incision. As in Pamphiliidae, the gregarious larvae feed in groups clustered in silken webs, although large-toothed sawflies feed and cluster on herbaceous plants. The webs of older larvae form tubes, with the larvae incorporating their feces in the tube wall, within which they hide. Since adults visit flowers for nectar they probably provide a pollination service.

DISTRIBUTION
Widely distributed across Eurasia, preferring steppe or Mediterranean-type climates

GENUS
Megalodontes

HABITATS
Warm, dry, open areas; rarest in the north and west of their range

SIZE
⁵/₁₆–⁹/₁₆ in (8–15 mm) body length

ACTIVITY
Diurnal for feeding

REPRODUCTION
Males seek out and mate with females on flowers. Copulation is prolonged

DIET
Phytophagous. Larvae feed only on herbaceous plants (Apiaceae), in contrast to the Pamphiliidae. Adults obtain nectar and pollen from flowers

WEB-SPINNING OR LEAF-ROLLING SAWFLIES

The family Pamphiliidae is confined to the Holarctic region. There are 2 extant subfamilies, Cephalciinae and Pamphiliinae, containing 8 living genera represented by 327 species, and 5 extinct genera, 3 of which are classified in the fossil subfamily Juralydinae.

As in the Megalodontesidae these are large sawflies with wide, flattened bodies. Their antennae are long and thin, tapering to points.

Eggs are laid in slits in the plant tissue. The gregarious larvae often feed in groups clustered in silken webs or silk tents on coniferous and other plants, with other species feeding within leaf rolls. Many species are gregarious in early larval instars, living in silk tubes, then solitary in later instars,

ABOVE | *Acantholyda erythrocephala,* the distinctive Pine False Webworm, or Red-headed Pine Sawfly; sawflies have often acquired different names in the pest control literature versus the recording community.

RIGHT | Pine Tree Web-spinner (*Acantholyda posticalis*). Although considered a forestry pest in parts of its range, in other areas, such as Britain, this is a declining species.

although some species are gregarious under silken webbing throughout their development. Many Pamphiliinae live within leaf rolls in later instars.

Pupation is in the soil and the winter is spent as a prepupa. Larvae of most species feed on trees, either conifers or angiosperms, while species that feed on herbaceous plants are restricted to Rosaceae.

DISTRIBUTION
Widely distributed across the temperate areas of Asia, Europe, and North America

GENERA
Acantholyda, Caenolyda, Cephalcia, Kelidoptera, Neurotoma, Onycholyda, Pamphilius, Pseudocephaleia

HABITATS
Woodland

SIZE
$^5/_{16}$–$^{11}/_{16}$ in (8–18 mm) body length

ACTIVITY
Diurnal

REPRODUCTION
Males approach females and mating lasts several minutes. Females can be aggressive, resisting further mating

attempts and at least in some
species exhibiting aggressive
egg-guarding behavior

DIET
Phytophagous. Larvae feed on
trees or herbaceous plants.
Adults do not feed

The subfamily Cephalciinae are pine feeders while
Pamphiliinae feed on angiosperms. Adults do not feed,
in contrast to Megalodontesidae.

Some species are considered to be pests of the forestry
industry, such as *Cephalcia lariciphila*, which can defoliate stands
of larch (*Larix*), and the Pine False Webworm (*Acantholyda
erythrocephala*), which attacks pines.

HORNTAILS AND WOODWASPS

This family is most common in the temperate northern hemisphere, but *Afrotremex* is an African genus and *Sirex noctilio* has spread to Africa, Australia, and South America. It comprises 2 extant subfamilies, Tremicinae (horntails) and Siricinae (woodwasps), containing 10 genera and 108 species. An additional nine genera and three subfamilies are only known by fossils, an ancient group of wasps that was more diverse during the Paleogene and Mesozoic eras.

Woodwasps develop within conifers and horntails feed on broad-leafed deciduous tree species. Horntails are so named for the hard spine-like projection at the end of the abdomen, which facilitates penetration of plant tissue in the process of egg-laying. They need toughened cuticle to oviposit into and emerge from wood, and have zinc-enriched ovipositor tips and mandibles to aid this. The Pigeon Horntail (*Tremex columba*) is a massive North American wasp reaching up to $1^{15}/_{16}$ in (50 mm) in body length.

Woodwasps carry symbiotic fungi in specialized "pouches," mycangia, on the abdomen. Adult woodwasp females lay eggs and fungal spores into stressed trees, providing rotten wood for larval feeding. *Xeris* species lack mycangia, taking advantage of wood pre-infested by woodwasps.

DISTRIBUTION
Species richness is centered in the northern hemisphere, but *Afrotremex* is present in the African tropics and *Sirex noctilio* has been introduced to several countries in the southern hemisphere

GENERA
Afrotremex, Eriotremex, Sirex, Siricosoma, Sirotremex, Teredon, Tremex, Urocerus, Xeris, Xoanon

HABITATS
Temperate and tropical forests

SIZE
$3/_8$–$1^{15}/_{16}$ in (10–50 mm) body length

ACTIVITY
Diurnal. Some species are reported to have mating swarms at the tops of trees or hills ("hilltopping")

REPRODUCTION
Females lay several eggs directly into the wood of dead or stressed trees,

with ovipositor valves working like saws to slice through wood fibers. The larvae are long-lived, for two to three years. They usually pupate just below the bark surface

DIET
Larvae are wood-borers, phytophagous or mycophagous, feeding on dead wood being broken down by commensal fungi

Sirex noctilio has become a major forestry pest and has been accidentally introduced to several countries in the southern hemisphere. Several other species are invasive, mainly in the Americas, where they are frequently intercepted in wood imports. *Sirex noctilio* is successfully controlled by a combination of parasitoid wasps (*Ibalia leucospoides* and species of Rhyssinae, Ichneumonidae) and parasitic nematodes (*Deladenus siricidicola*).

ARGID SAWFLIES

A rgids are nearly worldwide in distribution, but most prevalent in tropical regions. They are absent only from Antarctica, Madagascar, and New Zealand. They comprise 2 subfamilies, Arginae and Sterictiphorinae, containing 59 living genera, with over 900 species. One fossil genus has been described (two extant genera also have fossil species). The Zenarginae was recently raised to family status, and the other four subfamilies absorbed into the two current subfamilies based on phylogenetic analyses.

Argid sawflies are frequently brightly colored and conspicuous. A characteristic feature of Argidae (and a very few other sawflies) is that the antenna is reduced to three segments. Males of Sterictiphorinae have bifurcate (resembling tuning forks) antennae, whereas Arginae all have simple antennae.

This abundant family represents an important herbivory component in tropical ecosystems. Larvae may be gregarious, and a few exposed feeders are different colors on each side of their body, as camouflage for feeding on the edges of

DISTRIBUTION
Worldwide, with species richness centered in tropical and subtropical regions, although the huge genus *Arge* is well represented in temperate regions

GENERA
59 extant genera classified in 2 subfamilies: Arginae and Sterictiphorinae

HABITATS
Wide range, from arid desert environments to tropical forests. Most abundant in tropical areas

SIZE
$\frac{1}{8}$–$\frac{1}{2}$ in (3–12 mm) body length

ACTIVITY
Diurnal. Many argids are rather slow-moving, "bumbling" along

LEFT | An argid sawfly, *Arge ochropus*, with a typical bright color pattern for an argid, although the distantly related *Athalia* sawflies can look similar. Adult argids are often rather slow-moving.

RIGHT | The Poison Ivy Sawfly (*Arge humeralis*) with black-and-red warning colors. Note the antennal flagellum, reduced to one long segment.

BELOW | Larvae of the Large Rose Sawfly (*Arge pagana*), conspicuous and distasteful to many predators, although some specialized wasps are parasitoids of argid larvae.

REPRODUCTION
Females use the saw-like ovipositor to cut slits into host plant to lay eggs. Maternal care may occur with egg guarding. Larvae may be gregarious; Sterictiphorinae pupate in dense cocoon masses

DIET
Phytophagous, larvae feed externally on angiosperms. Some are leaf-miners with *Aproceros* and *Sterictiphora* leaving a distinct zig-zag pattern

leaves. Adults are flower visitors feeding on pollen and nectar, and consequently play a role in pollination services.

The family contains several pest species and a couple of beneficial species used as biocontrol agents. The Sumatran species *Cibdela janthina* has been successfully used to control the invasive Giant Bramble (*Rubus alceifolius*, Rosaceae) on the island of Réunion and is now also present in Mauritius. *Aproceros leucopoda* is a pest of elms (*Ulmus*), having invaded Europe from East Asia, and can completely defoliate the trees. The rose sawflies *Arge ochropus*, *Arge rosae*, and *Arge xanthogaster* are serious pests of both ornamental and wild roses, the gregarious larvae feeding on both flowers and leaves. Larvae of *Arge pullata*, a widespread species that can defoliate birch, are toxic to livestock, which frequently eat the larvae as they move to the ground to pupate.

FERN SAWFLIES

Blasticotomidae are restricted to Eurasia, occurring in temperate areas. This small family of sawflies comprises 2 living genera represented by only 12 species, and has 1 fossil genus.

As in Argidae and Zenargidae, most of the antenna is one enlarged flagellar segment, but this is followed by a tiny fourth antennal segment in Blasticotomidae.

The young larva chews out a cavity in the rachis of the fern foodplant, where it feeds on sap leaking into the hollow cell. The larva chews a hole at each end of the cavity it develops in, the bottom one for defecation and the other for breathing. The excrement is sugar-rich and attracts ants, their presence likely providing indirect protection for the larvae from parasitoids and predators. Pupation is within the stem, without a cocoon.

The most widespread species, *Blasticotoma filiceti*, is thelytokous (reproduces parthenogenetically); occasional introductions outside its native range, and in gardens, mean it can easily spread, as only one female needs to be transported.

BELOW | Female of a Fern Sawfly (*Blasticotoma filiceti*). Adults can be hard to find but the presence of larvae is often indicated by froth around a hole in a fern rachis, with ants attending.

DISTRIBUTION
Species are confined to temperate areas of Asia, with one species in Europe, and are rarely found in numbers

GENERA
Blasticotoma, Runaria

HABITATS
Boreal forests and wetlands, where ferns occur

SIZE
1/4–3/8 in (7–9 mm) body length

ACTIVITY
Diurnal

REPRODUCTION
Females lay eggs directly into the stem of a fern

DIET
Phytophagous. Larvae are stem-borers of ferns and can sometimes be detected by accumulations of froth outside their feeding hole

GIANT SAWFLIES

The family Cimbicidae is mostly a northern hemisphere family with one subfamily occurring in South America. It comprises 16 living genera represented by 182 species, and 5 extinct genera, in 4 subfamilies: Abiinae, Cimbicinae, Corynidinae, and Pachylostictinae.

Giant sawflies are a family of mostly large, fast-flying sawflies with characteristic clubbed antennae. Male fighting is known in some species and they can be highly territorial, using their massive mandibles and enlarged hind legs as weapons. This might be an evolutionary driver behind the very large body size of many species, as might mimicry, as some species mimic bumblebees (*Bombus*) or vespid wasps.

Cimbicinae feed mainly on trees, other subfamilies mainly on shrubs and herbs, although the biology of Pachylostictinae is very poorly known. Although the larvae are large and striking, cimbicids are usually found at low density. Larvae feed externally on foliage and pupate in cocoons constructed on the foodplant or in the soil.

RIGHT | Female of the European honeysuckle-feeding *Abia fasciata*. Larvae feed on a range of plants in the family Caprifoliaceae.

DISTRIBUTION
Species richness is centered in the northern hemisphere with Pachylostictinae restricted to South America

GENERA
16 genera

HABITATS
Wide range, from grassland to forests

SIZE
1/4–1 3/16 in (6–30 mm) body length

ACTIVITY
Diurnal. Adults often buzz when caught and have a strong defensive bite. Males aggregate at the top of hills

REPRODUCTION
Eggs are laid in incisions on foodplants

DIET
Phytophagous. Larvae are plant-feeders, all on angiosperms

CONIFER SAWFLIES

The family Diprionidae is restricted to the northern hemisphere, with species richness centered in the boreal regions of Asia, Europe, and North America, but some species extend south to the Oriental region and to North Africa.

Conifer sawflies comprise 11 living genera represented by 136 species, and 2 extinct genera in 2 subfamilies. Some species have gregarious larvae and can be serious forestry pests, such as the White Pine Sawfly (*Neodiprion pinetum*), which attacks the Eastern White Pine (*Pinus strobus*). Population explosions under favorable conditions can result in complete defoliation of forest trees resulting in the death of the tree, an impact attributing a major economic pest status to these species. *Neodiprion sertifer* is a notorious pest in North America, introduced from Europe, where massive defoliation of pines occurred before some control was achieved using parasitoid wasps.

Males of many species have extravagantly pectinate antennae, and these are attracted to pheromones produced by females. Larvae of solitary species are cryptically colored while gregarious larvae can be conspicuous and display synchronized movements when threatened. The very well-studied *Neodiprion sertifer* overwinters as an egg, unusual for Diprionidae; most species overwinter as prepupae in cocoons either in the ground or on the food tree.

LEFT | Female of the pine-feeding *Diprion similis*. In hot, dry years, there can be increased defoliation of pines as *Diprion* species can then manage two generations in a year.

DISTRIBUTION
Species richness is centered in Eurasia and North America

GENERA
Augomonoctenus, Diprion, Gilpinia, Macrodiprion, Microdiprion, Monoctenus, Neodiprion, Nesodiprion, Prionomein, Rhipidoctnus, Zadiprion

HABITATS
Temperate conifer forests

SIZE
$3/16$–$9/16$ in (5–14 mm) body length

ACTIVITY
Diurnal. Slow-flying, with males ranging much further than females

REPRODUCTION
Eggs are laid in slits on conifer needles

DIET
Phytophagous. Larvae feed on conifers, Diprioninae on Pinaceae and Monocteninae on Cupressaceae

FERN-TIP SAWFLY WASPS

The family Heptamelidae, only recently split from Tenthredinidae, is restricted in its native range to Eurasia. It comprises 2 living genera represented by 38 species, with no known extinct genera.

The larvae of *Heptamelus* are stem-borers whereas those of *Pseudoheptamelus* feed mostly externally. Eggs of both genera are laid at tips of the fern rachis, entirely within the plant tissue. Larvae of *Heptamelus* feed by boring downwards toward the base of the stem. Larvae of *Pseudoheptamelus* prefer to feed externally on freshly dead fern. The foodplants are known for very few species; European heptamelids seem to prefer Lady Fern (*Athyrium filix-femina*). The ovipositing females of *Pseudoheptamelus runari* make additional cuts with their ovipositors above the egg, which might help kill the fern frond, to the advantage of the larva. As with most other fern-feeding sawflies, larvae do not pupate in the soil nor form cocoons, but burrow into lower, dead sections of fern stem, or nearby dead wood to pupate.

BELOW | Female *Heptamelus dahlbomi*; males are unknown. Classified as tenthredinid sawflies until recently, this is a difficult family to recognize.

DISTRIBUTION
Species are most diverse in the Eastern Palearctic and Oriental regions, with a few species in Europe. *Heptamelus dahlbomi* has been accidentally introduced to North America

GENERA
Heptamelus, Pseudoheptamelus

HABITATS
Forests, wetlands (including dune slacks), gardens

SIZE
$3/16-1/4$ in (4–6 mm) body length

ACTIVITY
Diurnal

REPRODUCTION
One widespread species, *Heptamelus dahlbomi*, is thelytokous

DIET
Phytophagous on ferns

SPITFIRE SAWFLIES

The family Pergidae, with 60 living genera represented by 442 species, is the third-largest sawfly wasp family after the argids and torpedo sawflies (Tenthredinidae), being most species-rich in South America, followed by Australia. The family also occurs in eastern North America (the genus *Acordulecera*) and the Indonesian region (New Britain, Papua New Guinea, and Sulawesi). One species has been introduced to New Zealand.

RIGHT | Larval Australian pergid sawfly (*Trichorhachus* species) feeding on Common Smokebush (*Conospermum stoechadis*), and well disguised on its white fluffy foodplant.

BELOW | Adult Australian pergid sawfly (*Perga* species). *Perga* adults have extremely short antennae and the larvae are highly gregarious.

DISTRIBUTION
Species richness is centered in the southern hemisphere; it is the best-represented sawfly family in Australia and is well represented in South America

GENERA
60 genera contained in 12 subfamilies: Acordulecerinae, Conocoxinae, Loboceratinae, Parasyzygoniinae, Perginae, Pergulinae, Perreyiinae, Philomastiginae, Pterygophorinae, Pteryperginae, Styracotechyinae, Syzygoniinae

HABITATS
Wide range, from grassland to forests

SIZE
$1/8$–$3/8$ in (3–10 mm) body length

ACTIVITY
Most are diurnal, but some species are crepuscular or nocturnal

Males home in on receptive females by detecting pheromones released by the females. Pergids exhibit several unusual traits. Unusually within sawflies, some species exhibit maternal care, aggressively looking after their eggs and young larvae by making a buzzing sound. Sawfly adults are frequently carnivorous, so this is not an empty threat, and some pergids are known to be highly toxic. Another unusual feature, found in *Perreyia flavipes*, is thanatosis—that is to say, they "play dead."

Larvae feed mostly on leaves (commonly eucalypts in Australia) or are borers attacking new shoots or mining within the leaves themselves. Others only feed on decomposing plant matter. Several species are of economic significance, as the gregarious larvae can completely devour all the leaves on the host plant, which can be crops such as guavas or potatoes in South America, or forest trees. The South American *Perreyia flavipes* has gregarious and highly toxic larvae that are sometimes known as "pig killers," responsible for many livestock deaths in Brazil. Others are beneficial, helping to control invasive plant species such as the Brazilian Peppertree (*Schinus terebinthifolia*) and the Peruvian Peppertree (*Schinus molle*).

REPRODUCTION
Fertilized females usually lay their eggs by inserting them directly into the host plant tissue, but sometimes lay eggs on or in the soil

DIET
Extremely varied, most are phytophagous (plant-feeders) although some specialize on aquatic ferns, and others on fungi

TORPEDO SAWFLIES

ABOVE | *Rhogogaster viridis*; as with many other bright green insects, the green pigment fades after death.

OPPOSITE | *Tenthredo* species, either *T. arcuata* or a close relative; another mimic of stinging wasps. *Tenthredo* species are active, relatively large, and eat other insects.

The Tenthredinidae are by far the largest family of sawflies, with 400 genera. Torpedo sawflies occur worldwide but are a particularly conspicuous part of the northern temperate fauna. The number of subfamilies recognized in Tenthredinidae has been unsettled but we follow recent phylogenetic studies that agree on six subfamilies: Allantinae, Athaliinae, Blennocampinae, Nematinae, Selandriinae, and Tenthredininae.

DISTRIBUTION
Species richness is centered in the temperate areas of North America and Eurasia, but members of most subfamilies range far south into Africa and South America, and Selandriinae is more diverse in the tropics

GENERA
400 genera in 6 subfamilies: Allantinae, Athaliinae, Blennocampinae, Nematinae, Selandriinae, Tenthredininae

HABITATS
Wide range, from more arid environments to tropical forests, but most species-rich in temperate areas

SIZE
$3/16$–$3/4$ in (5–20 mm) body length

ACTIVITY
Diurnal, although some species appear in light traps and might be partly crepuscular

REPRODUCTION
Females use their saw-like ovipositor to
cut slits into plants for egg-laying

DIET
Phytophagous, mainly on angiosperms
but with significant numbers of species
feeding on ferns and monocots, and a
few on conifers

In line with the majority of recent phylogenetic
studies, we recognize Athaliinae as a separate
subfamily, as it has usually been found to be the sister
to the rest of Tenthredinidae, or could even be a
separate family within Tenthredinoidea. The former
subfamily Heterarthrinae has recently been split,
with its genera reassigned to Blennocampinae.

While many of the smaller species tend to be monochromatic and difficult to identify on external features, larger species of Tenthredininae can be very colorful, with some vivid green and orange species. Athaliinae species have distinctive black- and yellow-striped legs and a mainly orange abdomen.

In sub-Arctic and Arctic areas, species in the Tenthredinidae family are particularly important components of the plant-feeding insect community. Tenthredinids are very important hosts for several groups of Ichneumonidae and a few other groups of parasitoids.

Quite a few species bring attention to themselves as agricultural and ornamental plant pests, such as the Pear Slug (*Caliroa cerasi*), which feeds on ornamentals such as roses and deciduous fruit trees, or the Solomon's Seal Sawfly (*Phymatocera aterrima*), which defoliates Polygonatum ornamentals. The European Turnip Sawfly (*Athalia rosae*) has caused

ABOVE | Mating pair of the Turnip Sawfly (*Athalia rosae*). Very recently, the Athaliidae have been recognized as a family, separate from Tenthredinidae.

BELOW | The strange-looking "slugworm" larva of *Caliroa cerasi* are covered in mucous and feed on the leaves of various trees and shrubs.

RIGHT | Female *Strongylogaster multifasciata* ovipositing in a frond of bracken; the larvae feed externally on the foliage.

extensive damage to crops in Northern Europe but has become much less common. However, *Athalia rosae* and sometimes other *Athalia* have regular population explosions, when they can arrive in swarms in coastal areas.

Larvae of most species feed externally on leaves, but some are gall-formers, stem-borers, or leaf-miners. Larvae of Allantinae and Athaliinae are external feeders, the former on a variety of trees and shrubs (with a preference for Rosaceae), the latter on herbaceous plants and particularly Brassicaceae. Blennocampinae is a diverse subfamily containing leaf-miners (fenusines, heterarthrines), leaf skeletonizers (caliroines), borers in roots and twigs, and many external feeders. While angiosperms are preferred by most, with many larvae on trees, a few species eat monocots. The more than 1,000 species of Nematinae feed mainly on angiosperm trees, but with radiations onto various other plants.

Gall-forming and leaf-rolling are particularly prevalent in species feeding on *Salix*, and many of the nematines that feed on willows will only feed on one or very few species. With willows being species-rich in the far north, this has resulted in huge numbers of nematine species, particularly in the genera *Euura* and *Pristiphora*. Selandriinae are largely fern feeders, including *Strongylogaster*, which has been used to try and control Bracken (*Pteridium aquilinum*). Some species eat monocots, including the tribe Dolerini, often common in grassland. Adults of many tenthredinids feed on nectar and pollen, and larger tenthredinids, such as the genus *Tenthredo*, which are predators of other insects, often hunt on flowers.

Most temperate species only have a single generation every year and adults are most conspicuous in spring and early summer, but some (e.g., *Athalia*) produce multiple generations in a year. Larvae of solitary species tend to be cryptic, but gregarious species can have strikingly conspicuous larvae. The larvae pupate in the ground or in a protected situation such as in their galls, stems, leaf-mines, etc., and survive adverse cold season conditions in this phase of their life cycle.

CYPRESS PINE SAWFLIES

The family Zenargidae was only recognized as distinct in 2021. The single included species, *Zenarge turneri*, is endemic to Eastern Australia and had previously been classified in the family Argidae.

Large areas of *Callitris* and *Cupressus* have been defoliated by *Z. turneri*, with younger trees preferred.

Emergence of adults requires a certain amount of winter rainfall and moderate temperatures at other times. As a consequence, *Zenarge* can spend up to six years as a prepupa in the cocoon, waiting for the right conditions.

Larvae of *Z. turneri* feed on gymnosperms, including cypress pine, native to Australia, as well as introduced conifers, which contrasts with the closely related families Argidae and Pergidae, mostly angiosperm feeders.

ABOVE | Cypress Pine Sawfly (*Zenarge turneri*). As with argid sawflies, the antennal flagellum is reduced to one large segment.

DISTRIBUTION
Eastern Australia

GENUS
Zenarge

HABITATS
Coastal areas and highlands of eastern Australia, where *Callitris* (cypress pine) and introduced *Cupressus* (cypress) occur

SIZE
$5/16–3/8$ in (8–10 mm) body length

ACTIVITY
Adults are diurnal and emerge at various times of the year, but particularly the fall. Little is known about their behavior

REPRODUCTION
Eggs are laid singly on shoots of gymnosperms and larvae are not gregarious

DIET
Phytophagous

LONG-NECKED WOODWASPS

The family Xiphydriidae is represented worldwide on most continents, except for sub-Saharan Africa. It comprises 27 living genera represented by 153 species contained in 2 subfamilies, Derecyrtinae and Xiphydriinae, and 2 extinct genera.

Long-necked woodwasps are rarely collected but some species can be found to be reasonably common when their feeding habits are known.

Females inject fungal spores during the process of egg-laying. The larvae require the presence of these symbiotic fungi that subsequently grow in the tunnel they bore in the host trees for their development. Females oviposit in wood using their conspicuously exserted ovipositors.

Several eggs are laid in one drill hole and the larvae then bore fairly shallow tunnels in the wood. As in siricid woodwasps, larvae have a stiff structure at the rear end that might be used by the larva as a prop while rasping at hard substrate. Larvae are wood-borers in dead or decaying branches of deciduous trees in the families Aceraceae, Betulaceae, Salicaceae, and Ulmaceae. Feeding is essentially on the fungal hyphae breaking down the wood.

BELOW | Female Alder Woodwasp (*Xiphydria camelus*). The natural history of this species and its parasitoids was documented in a groundbreaking 1961 film by Gerald Thompson.

DISTRIBUTION
Xiphydriinae are worldwide in distribution with the exception of sub-Saharan Africa

GENERA
27 genera

HABITATS
Forests and woodland

SIZE
$1/2$–$3/4$ in (12–20 mm) body length

ACTIVITY
Diurnal. As with some siricids and cimbicids, males have been found in swarms at the tops of hills, where presumably females find them

REPRODUCTION
Females drill holes in wood with their ovipositor and lay eggs in the hole

DIET
Phytophagous/mycophagous

WEIRD-ANTENNAE SAWFLIES

The family Xyelidae is restricted to the northern hemisphere, mostly occurring in temperate areas. It comprises 5 living genera represented by 63 species, classed in 2 subfamilies: Macroxyelinae and Xyelinae. There are also 53 extinct genera.

This is an ancient radiation of sawflies, with species richness best represented in the fossil record (120 species). Some of the oldest Hymenoptera fossils dating back to the Triassic (245–208 MYA) belong to this family.

Although they attack tree species of economic importance, most species are not major pests. One species, *Megaxyela major*, is regarded as a pest of Pecan plantations in the USA.

OPPOSITE | A North American *Xyela* species. *Xyela* are relatively diverse in North America but species are difficult to identify.

BELOW | Female *Xyela* species. The long ovipositor is used to lay eggs between the scales of pine cones.

DISTRIBUTION
Species richness is centered in the boreal temperate regions, but some species extend to the Oriental region

GENERA
Macroxyelinae: *Macroxyela*, *Megaxyela*, *Xyelecia*; Xyelinae: *Pleroneura*, *Xyela*

HABITATS
Temperate forests

SIZE
$1/16$–$9/16$ in (2–15 mm) body length

ACTIVITY
Diurnal. Adults can sometimes be swept in numbers, especially around pines

REPRODUCTION
Females lay eggs in pinecones or pine leaf buds

DIET
Phytophagous. Larvae of species
in the subfamily Macroxyelinae
feed on leaves of deciduous trees,
commonly species in the walnut family
(Juglandaceae), or on elms. Those in
the subfamily Xyelinae feed on pollen
(or within the developing buds) of
conifers, commonly firs, pines,
or spruces

The large maxillary palps present in species of Xyelinae
have evolved to facilitate extracting pollen from flowers.

Reproductive strategies differ markedly between genera.
Females of *Xyela* oviposit in staminate (pollen-producing)
pinecones. *Xyelecia* and *Pleroneura* oviposit in buds and young
shoots of pines. *Macroxyela* stick their eggs to the leaf surface
after folding the leaf using their long hind legs. Larvae of
xyelids burrow into the ground to pupate within a silk cocoon
spun within an earthen cell.

Most species are small, but *Megaxyela* are around ⁹⁄₁₆ in
(14–15 mm) long and with very long hind legs.

APOCRITA (PARASITICA)
PARASITOID WASPS

NARROW-WAISTED WASPS (APOCRITA)

Historically this group of wasps was ranked at suborder level as a sister group (Apocrita) to the Symphyta, but the apocritan lineages have evolved from within the Symphyta, so the two groups cannot be considered as monophyletic suborders. The narrow-waisted wasps contain two groups, Aculeata and Parasitica, which were each historically regarded as infraorders in the classification, but the Aculeata have evolved from within the Parasitica, which means that they also cannot be considered to be sister groups at the same level of classification (see the phylogenetic tree on page 30).

The Apocrita are defined by the possession of a "wasp waist" that is absent in the symphytans. The waist is formed by a constriction between the first and second abdominal segments. The first abdominal segment is named the propodeum and is fused to, and effectively functions as part of the thorax. Consequently, these two body parts are referred to as the mesosoma (functional thorax) and the metasoma (functional abdomen).

The second, or second and third abdominal segments (therefore first or first and second metasomal segments), often form a petiole, which is a thinner segment than the remaining metasoma.

LEFT | A female parasitoid fig wasp, *Apocrypta guineensis* (Pteromalidae) ovipositing through the fig wall to reach the host fig wasp larvae already developing in the galled florets within the central fig cavity of *Ficus sur*.

Many ants have two petiolar segments. The evolution of this narrowed waist section, often with a distinct petiole, was an adaptive morphological response to enable improved mobility of the metasoma and more accurate positioning of the ovipositor for the parasitoid way of life.

We here maintain these two informal groupings Parasitica and Aculeata for ease of reference to historical classifications.

PARASITOID WASPS (PARASITICA)

The Parasitica have historically been classified as an infraorder of the Apocrita, but the Aculeata lineage evolved from parasitoid wasp ancestors, so the Parasitica cannot be a monophyletic group classified at the same level (infraorder) as the Aculeata. The Aculeata also contain many species, including several families, that are parasitoids, negating the circumscription of a Parasitica group based on their lifestyle strategy. This assemblage (the Hymenoptera, which are not sawflies or woodwasps, and not aculeates) comprises 12 superfamilies, of very different morphologies.

The majority of wasp species belong here, with their elevated species richness and diversity probably having been driven by their parasitoid existence. Most species of insects and spiders (plus a few other arachnids) are attacked by one—but often more than one—species of parasitoid wasp. Although parasitoid wasps are often assumed to be frequently host-specific, this is not usually the case. Instead, a range of hosts is usually attacked by any given parasitoid species, although this range is tightly circumscribed by factors such as host taxonomy, host niche, seasonality, etc. Some are, of course, absolutely host-specific. The females lay their eggs on or inside the host, with the larvae consuming the host.

Parasitoids are differentiated from parasites as parasitoids kill their host, with each parasitoid consuming one (or part of one) host, so a mix of parasite and predator, whereas parasites (such as ticks feeding on vertebrates) do not usually kill their host, except sometimes indirectly via pathogen transmission.

Many parasitoid wasps are small, averaging $1/32–1/16$ in (1–2 mm) in length, although some can be up to several centimeters long. The venoms of parasitoid wasps are often chemically and physiologically complex and some groups of Ichneumonoidea have co-opted viruses, all as adaptations to overcome hosts and their immune responses. Some of the larger Darwin wasps with short ovipositors (e.g., Ophioninae, some Ichneumoninae, *Netelia*) have fairly painful stings that can be used defensively. A few lineages of parasitoid wasps have lost their parasitoid lifestyle and are instead phytophagous (most notable in the gall wasps) or predatory, such as various Darwin wasps that consume spider egg sacs.

LEFT | A female *Gasteruption jaculator* ovipositing into a solitary bee nest.

TRIGONALYID WASPS

The family Trigonalyidae has a worldwide distribution and includes 2 subfamilies, Orthogonalyinae and Trigonalyinae, with 16 extant genera represented by around 120 species, and 4 extinct genera. The Cretaceous fossil genus *Maimetsha* (Maimetshidae) was originally believed to have affinities with the Megalyridae but is now considered to be related to the Trigonalyidae, and Maimetshidae is included in the superfamily Trigonalyoidea.

The elongate species mimic Darwin wasps (Ichneumonidae) in appearance and the more robust species mimic social wasps. Trigonalyids can be overlooked as aculeate wasps with very complete wing venation, but the antennae have more segments (usually 18–28), the mandibles are large and usually asymmetrical, the tarsal segments have plantar lobes (otherwise not found in the Apocrita), and the female antennae have unique white, scale-like setae.

LEFT | The North American *Taeniogonalos gundlachii.*

ABOVE RIGHT | An African species, *Trigonalys natalensis.*

DISTRIBUTION
Worldwide, except for the Arctic regions and New Zealand. Most species occur in the tropics

GENERA
Afrigonalys, Bakeronymus, Bareogonalos, Ischnogonalos, Jezonogonalos, Lycogaster, Mimelogonalos, Nomadina, Orthogonalys, Pseudogonalos, Pseudonomadina, Seminota, Taeniogonalos, Teranishia, Trigonalys, Xanthogonalos

HABITATS
Forests, woodland, grassland

SIZE
1/8–9/16 in (3–15 mm) body length with much variation within species, depending on the size of the host; some elongate, others stout-bodied

ACTIVITY
Diurnal. Adults are usually rarely encountered but can be more common at higher altitudes in Asia and can be

These wasps are usually highly specialized hyperparasitoids of ichneumonoid wasp or tachinid fly parasitoids in plant-feeding caterpillars, or sometimes parasitoids of social wasps (Vespidae) when fed on parasitized caterpillars. However, at least one Australian species is known to be a primary parasitoid of pergid sawflies. The biology of trigonalyids is unusual for parasitoid wasps, because egg-laying does not occur directly into, or onto, the host.

Trigonalyids have an unusual life history as indirect hyperparasitoids. Females lay huge numbers of tiny eggs on or in slits in leaves,

which are then consumed by butterfly or moth caterpillars, or sawfly larvae. The eggs hatch within the caterpillar, and the larvae chew through the gut and into the body of the caterpillar, where they attack other parasitoid wasp larvae (Ichneumonidae or Braconidae) or parasitoid fly larvae (Tachinidae) already inside the host, thus developing as hyperparasitoids. If the plant-feeding caterpillar is fed to a social or potter wasp (Vespidae) larva, the trigonalyid then attacks the vespid larva developing inside the social wasp nest. Trigonalyids complete their development in the host pupa.

There are a few other groups of parasitoid wasps who lay eggs on foliage, such as the chalcidoid family Eucharitidae and the Darwin wasp genus *Euceros*, but in these cases the larvae hatch before contacting hosts (or carriers). Note that Trigonalidae and Trigonalyidae (and equivalent family-group names) are both in use, based on different interpretations of the derivations of the genus name *Trigonalys*.

locally more numerous, such as in cacao plantations in Brazil. They are typically found in forests. Usually only a tiny proportion of reared caterpillars produce trigonalyids

REPRODUCTION
Females lay eggs on or in slits in leaves

DIET
Usually highly specialized hyperparasitoids of other parasitoid wasp or fly larvae

MEGALYRID WASPS

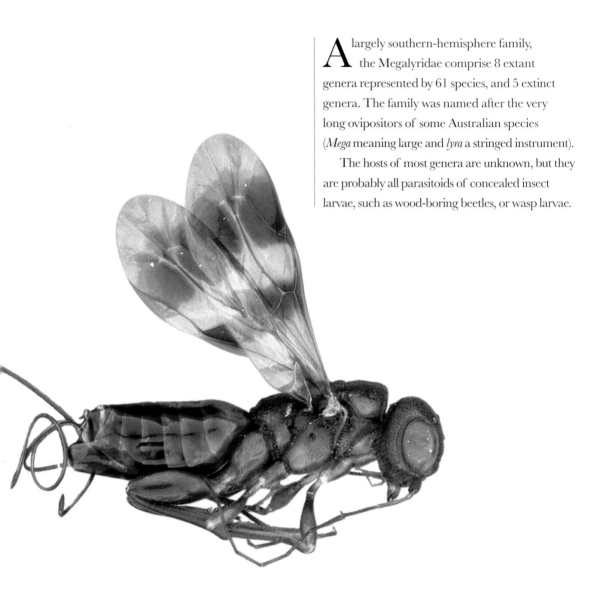

A largely southern-hemisphere family, the Megalyridae comprise 8 extant genera represented by 61 species, and 5 extinct genera. The family was named after the very long ovipositors of some Australian species (*Mega* meaning large and *lyra* a stringed instrument).

The hosts of most genera are unknown, but they are probably all parasitoids of concealed insect larvae, such as wood-boring beetles, or wasp larvae.

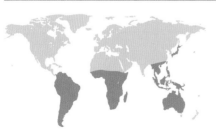

DISTRIBUTION
Mostly southern hemisphere, highest species richness in Australia and Madagascar, one species in Japan. Baltic amber and Cretaceous fossils demonstrate formerly more widespread distribution

GENERA
Carminator, Cryptalyra, Dinapsis, Ettchellsia, Megalyra, Megalyridea, Neodinapsis, Rigel

HABITATS
Tropical or temperate forests, woodland, dry scrub, fynbos

SIZE
$^1/_8$–$^7/_8$ in (3–22 mm) body length, excluding the ovipositor, which can be as long as $3^1/_8$ in (80 mm)

ACTIVITY
Diurnal. Adults are rarely encountered

The long ovipositors of the more conspicuous species are
not used for drilling; instead, the ovipositor is inserted into
pre-existing tunnels or holes to reach host larvae.

Little is known about the biology of most species, but
a common Australian species, *Megalyra fasciipennis*, attacks
large larvae of wood-boring beetles (Cerambycidae) and
has been introduced to several countries to control pest
longicorn beetle larvae (*Phoracantha*) developing in commercial
Eucalyptus plantations. The female inserts her long ovipositor
through the frass plug blocking the tunnel that the beetle
larva has bored while feeding, laying an egg on the host beetle
pupa. Another Australian species (*Megalyra troglodytes*) attacks
the larvae of mud-nesting crabronid wasps.

Some distinctive characteristics of Megalyridae include
the presence of subantennal grooves and highly reduced
hindwing venation.

REPRODUCTION
Little is known about the biology.
The female lays an egg on the host
insect developing inside wood or
mud nests using her long ovipositor

DIET
The parasitoid larva feeds on the
host larvae of wood-boring beetles
or wasps

CROWN WASPS

The Stephanidae contain 11 extant genera represented by 345 species, and 4 extinct genera.

They are easy to recognize by the distinctive crown of teeth on top of the globular head, and hind legs with bulbous first two segments, the second with large teeth on the underside. The crown of teeth on the head is shared with only one other wasp family, the Orussidae, and probably evolved to facilitate emergence of the adult wasp from deep inside wood.

OPPOSITE | A male of a *Foenatopus* species.

BELOW | The North American *Megischus bicolor* ovipositing into a host beetle larva developing inside a tree branch.

DISTRIBUTION
Worldwide, with most species occurring in tropical or subtropical forests

GENERA
Afromegischus, Comnatopus, Foenatopus, Hemistephanus, Madegafoenus, Megischus, Parastephanellus, Profoenatopus, Pseudomegischus, Schlettererius, Stephanus

HABITATS
Tropical or temperate forests, woodland, grassland, scrub

SIZE
$3/16$–$1 3/8$ in (3.5–35 mm) body length

ACTIVITY
Diurnal. Adults are rarely encountered

Females drill their ovipositor valves through wood to reach the targeted host larvae developing inside branches. They lay a single egg on the host larva. On hatching, the wasp larva develops as an external parasitoid. Species are parasitoids of mostly wood-boring beetle larvae (commonly Buprestidae and Cerambycidae), or occasionally sawflies or bees. Based on adaptations of the fore and hind tibiae, stephanids are presumed to have the ability to detect vibrations made by wood-boring larvae.

Adults live for three to four weeks. Individuals belonging to the same species can vary greatly in size, with the largest being twice the size of the smallest. This size variation is related to the size of the host insect larva that the crown wasp larva developed on.

Schlettererius cinctipes has been introduced to Tasmania from California as a biocontrol agent of the Sirex Woodwasp (*Sirex noctilio*, Siricidae).

REPRODUCTION
Females lay an egg directly onto host larvae developing inside branches

DIET
The parasitoid larvae feed on beetle, sawfly, or bee larvae

CERAPHRONID WASPS

OPPOSITE | *Ceraphron* searching for hosts to parasitize.

BELOW | A *Ceraphron* female— the twin fore-tibial spurs are visible on the front leg.

The family contains 16 genera represented by 304 described species and is widespread throughout the world. Most species are still undescribed.

These wasps are mostly commonly found in leaf litter and soil, with many being wingless, an adaptation for living in this environment.

Uniquely within Apocrita, Ceraphronoidea have two fore tibial spurs. A curious structure that is only found in Ceraphronidae, called the Waterston's evaporatorium, is present on the last (sixth) segment of the metasoma. This is

DISTRIBUTION
Worldwide

GENERA
Abacoceraphron, Aphanogmus, Ceraphron, Cyoceraphron, Donadiola, Ecitonetes, Elysoceraphron, Gnathoceraphron, Homaloceraphron, Kenitoceraphron, Masner, Microceraphron, Pteroceraphron, Retasus, Synarsis, Trassedia

HABITATS
Wide range of habitats, from forests to tundra

SIZE
$1/64$–$1/16$ in (0.5–2 mm) body length

ACTIVITY
Probably mostly diurnal

a concave area, often covered by cells in a honeycomb-like arrangement, and appears to be associated with glands whose secretions may be involved in courtship or defense.

This family is one of the wasp groups associated with the leaf litter and soil surface environment. Species are parasitoids of a range of host insects, commonly flies, but also moths and butterflies, beetles, true bugs, thrips, and lacewings (wax flies or dustywings, Coniopterygidae); one species has been reared from caddisfly pupae.

The developmental biology of ceraphronid wasps is interesting and under-studied; many species are known to be ectoparasitoids but some are endoparasitoids, and species of both strategies can be found in the same genus (*Aphanogmus*). While most are primary parasitoids, some species are hyperparasitoids of wasp primary parasitoids (Braconidae and Chalcidoidea) attacking scale insects or aphids.

This variation in life history is not reflected in morphology, with Ceraphronidae being one of the wasp families of the most uniform appearance.

REPRODUCTION
Females lay eggs directly on or into the host, wherein the larva hatches and consumes the host or primary parasitoid's body contents

DIET
The parasitoid wasp larvae feed on the larvae, pupae, or immature stages of a range of host insects

MEGASPILID WASPS

The family includes 13 genera represented by 299 species in 2 subfamilies: Lagynodinae and Megaspilinae.

Unlike the predominantly terrestrial, ground-surface-inhabiting sister family Ceraphronidae, many megaspilids are more arboreal with good flight capabilities, although several species are wingless or short-winged, and hence occupy a similar terrestrial habitat.

The sexes are dimorphic in the Lagynodinae, with females wingless and males winged and appearing completely different.

Species are external primary parasitoids of fly pupae (Diptera), scale insects (Hemiptera), lacewings (Neuroptera), or in one known case, snow fleas (Mecoptera: Boreidae). While the biology of most species is poorly known, some are well studied as hyperparasitoids of other wasps (Braconidae) through aphids, where *Dendrocerus* parasitoids can be reared from hyperparasitoids in a Russian doll–type situation, up to four levels of hyperparasitoids.

LEFT | A male *Dendrocerus* species. Males have branched antennae, an adaptation to increase surface area and hence the number of chemical receptors for olfactory detection of female pheromones for mate location.

ABOVE RIGHT | *Aulacus burquei*, a parasitoid of *Xiphydria* woodwasp species (Xiphydriidae).

BELOW RIGHT | A *Pristaulacus* female ovipositing into a host wood-boring beetle larva feeding inside a tree trunk.

DISTRIBUTION
Worldwide

GENERA
Archisynarsis, Aetholagynodes, Conostigmus, Creator, Dendrocerus, Holophleps, Lagynodes, Megaspilus, Platyceraphron, Prolagynodes, Trassedia, Trichosteresis, Typhlolagynodes

HABITATS
Wide range of habitats, from forests to tundra

SIZE
$1/16$–$1/8$ in (1.5–3 mm) body length

ACTIVITY
Diurnal

REPRODUCTION
Females lay eggs directly on the host, wherein the larva hatches and consumes the host or primary parasitoid's body contents

DIET
The parasitoid larvae feed externally

AULACID WASPS

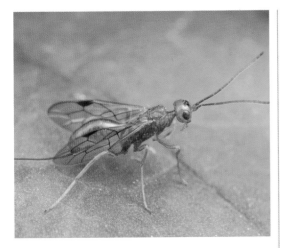

Aulacid wasps are a worldwide but thinly distributed family with 2 genera, *Aulacus* and *Pristaulacus*, represented by about 260 living species that are usually uncommon, although some species may be locally abundant during logging operations. *Panaulix* has been synonymized with *Pristaulacus*. The family is well represented in the fossil record, with an abundance of species known from the Mesozoic era contained in another 9 genera and 4 subfamilies.

These wasps are more prevalent in forested areas where there are decaying trees infested by their host wood-boring larvae. They are internal parasitoids of the larvae of wood-boring beetles (mainly Buprestidae and Cerambycidae, but also Bostrichidae, Cleridae, and Scolytidae) or woodwasps (Xiphydriidae). Females lay eggs directly into the host, drilling their ovipositor through wood to reach the host larva, wherein the larva hatches and consumes the host's body contents.

Life histories and host records are mostly sketchy but the European *Aulacus striatus* has been well studied and detailed films were made of its biology together with other parasitoids of the Alder Woodwasp (*Xiphydria camelus*).

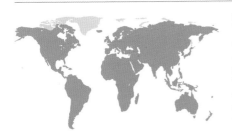

DISTRIBUTION
Worldwide

GENERA
Aulacus, Pristaulacus

HABITATS
Wide range of habitats, but more common in forested or woodland areas where their wood-boring hosts are present

SIZE
5/16–1 in (8–25 mm) body length

ACTIVITY
Diurnal. Adults are rarely encountered

REPRODUCTION
Females lay eggs directly into the host

DIET
Feed on larvae of wood-boring beetles or woodwasps

ENSIGN WASPS

ABOVE | *Evania appendigaster,* a predator of the ubiquitous American and Oriental cockroach pest species.

OPPOSITE | *Prosevania* species, a predator of concealed cockroach egg cases, the larvae feeding on the eggs and pupating inside the ootheca.

The family Evaniidae is known by a variety of common names alluding to the shape of the small flag-like laterally compressed metasoma attached very high on the propodeum by a petiole, or to the host of these wasps: ensign wasps, nightshade wasps, hatchet wasps, or cockroach egg parasitoid wasps.

Ensign wasps have a worldwide distribution, with 21 genera and about 400 described species, as well as 12 fossil genera, though many more undescribed species are known.

Evaniids oviposit a single egg into the concealed egg cases (oothecae) of cockroaches, and the larvae feed on the eggs and pupate inside the ootheca.

DISTRIBUTION
Worldwide

GENERA
Acanthinevania, Afrevania, Alobevania, Brachevania, Brachygaster, Decevania, Evania, Evaniella, Evaniscus, Hyptia, Micrevania, Paraevania, Papatuka, Prosevania, Rothevania, Semaeomyia, Szepligetella, Thaumatevania, Trissevania, Vernevania, Zeuxevania

HABITATS
A wide range of habitats, from forests to tundra, but most abundant in tropical forests

SIZE
3/16–9/16 in (5–15 mm) body length

ACTIVITY
Diurnal

REPRODUCTION
Evaniids oviposit into the concealed egg cases (ootheca) of cockroaches

They can lay an egg into oothecae that are still being carried around attached to the abdomen of a female cockroach, but many species will only lay an egg into the ootheca after it has been dropped. Cockroach egg cases have a tough integument that requires sustained effort by the female ensign wasp

DIET
Parasitoids of cockroach eggs, larvae feeding on the internal egg or developing nymph tissue. Adults visit flowers to obtain nectar

to drill through, a process that can take as long as half an hour. Pupation occurs inside the ootheca, and no cocoon is formed. Adults use their mandibles to cut their way out of the ootheca and can live for a couple of weeks.

The worldwide species *Evania appendigaster* may be common in dwellings because it parasitizes the ubiquitous American and Oriental cockroaches (*Periplaneta americana, Blatta orientalis*), which favor human habitation. Perhaps unfortunately, they do not attain population sizes that are efficient in controlling these pest species, so are of limited use in biocontrol programs.

CARROT WASPS

Carrot wasps are a globally distributed family represented by 6 genera with about 603 described species (most belonging to the genus *Gasteruption* in 2 subfamilies, Gasteruptiinae (*Gasteruption, Plutofoenus, Spinolafoenus, Trilobitofoenus*) and Hyptiogastrinae (*Hyptiogaster* and *Pseudofoenus*), and many more undescribed species.

Adults congregate in sleeping aggregations within sheltered environments such as inside a bush and are common visitors at flowers, particularly plant species in the carrot family, hence their common name.

LEFT | Carrot wasps have a characteristic "neck" that is formed by the modified propleuron.

ABOVE RIGHT | *Gasteruption* female at rest; her long ovipositor is used to probe deeply into solitary bee or wasp nests to reach the host larva for egg-laying.

DISTRIBUTION
Worldwide, but more species present in tropical than temperate areas

GENERA
Gasteruption, Hyptiogaster, Plutofoenus, Pseudofoenus, Spinolafoenus, Trilobitofoenus

HABITATS
Wide range of habitats, from forests to tundra

SIZE
$^1/_2$–$1^9/_{16}$ in (13–40 mm) body length

ACTIVITY
Diurnal. Adults may be observed visiting flowers for nectar, or around buildings where they search for host bee or wasp nests that are commonly attached to walls or under roof eaves

REPRODUCTION
Females lay an egg into a cell of the host nest containing an egg or developing larva

Carrot wasps feed on both the host larvae and food provision of cavity-nesting solitary bees (Apidae, Colletidae, Halictidae, and Megachilidae) and less commonly wasps (Crabronidae, Vespidae, and Sphecidae). The nests of the host can be situated in cavities in a variety of substrates including twigs, stems, galls, wood, or vertical soil surfaces.

Females lay an egg into a cell of the host nest containing an egg or developing larva. On hatching,

the carrot wasp larva usually eats the host egg or larva first before consuming the brood provision; some species wait for the host larva to complete development before consuming it. The bee *Hylaeus pectoralis*, which often nests in old Frit Fly (*Oscinella frit*) galls made in Phragmites reeds in Europe, is attacked by at least five *Gasteruption* species.

Adult carrot wasps take nectar at flowers, especially those of species in the carrot family (Apiaceae) in the northern hemisphere and Myrtaceae in Australia. Some species can be common in gardens, utilizing "bee hotels."

A distinctive feature of carrot wasps is the grossly enlarged hind tibia, which is full of fat. It is thought that this might act to amplify vibrations when searching for hosts, and also enable the hind legs to act as stabilizers when they are engaged in their dynamic flight, characterized by frequent hovering.

DIET
Predators and inquilines. The wasp larva consumes the host egg or larva as well as the pollen mass (in the case of bees), or paralyzed invertebrates (in the case of wasps) provisioned for the host larva by the adult female bee or wasp

FALSE FAIRY WASPS

This rarely collected, archaic, and unusual family contains 3 genera with 19 species of tiny wasps, and 2 extinct genera. Previously placed in the Chalcidoidea, Mymarommatidae is the only family with living species in the superfamily Mymarommatoidea, which also contains two families represented only in the fossil record: Gallorommatidae and Alvarommatidae.

The anatomy of mymarommatids is unique, with two tubular petiole segments, spoon-shaped forewings with a net-like surface sculpture, hindwings reduced to small forked projections, and the back of the head with pleats, looking like a bellows.

Only one species has been reared, in Hawaii, and this is a solitary parasitoid of bark lice (Psocodea: Lepidopsocidae) eggs laid on the trunk of a fig tree.

BELOW | An undescribed species of *Mymaromma* from Central Africa.

DISTRIBUTION
Worldwide

GENERA
Mymaromma, Mymaromella, Zealaromma

HABITATS
A wide range of habitats, from forests to tundra

SIZE
Average $1/64$ in (0.5 mm) body length

ACTIVITY
Diurnal. Adults are rarely encountered but can be swept or Malaise-trapped. Adults of *Mymaromma menehune* have been found walking on branches where their host's eggs are laid

REPRODUCTION
Where known, a single egg is laid in a host bark louse egg

DIET
Parasitoid feeding on egg contents

AUSTRONIID WASPS

Austroniid wasps are an endemic Australian family containing only a single genus with 3 species (*Austronia nigricula*, *A. nitida*, and *A. rubrithorax*).

They occur in the temperate forests of southeast Australia and Tasmania.

Their biology is unknown.

BELOW | Female of an *Austronia* species; all are rare wasps with a restricted distribution.

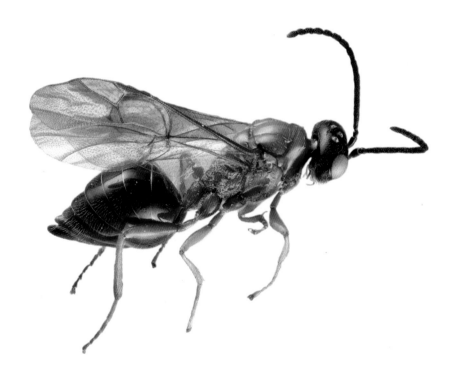

DISTRIBUTION
Australia, where they only occur in the forests of southeastern Australia and Tasmania

GENUS
Austronia

HABITATS
Temperate forests

SIZE
$^3/_{16}$–$^1/_4$ in (4–6 mm) body length

ACTIVITY
Adults are rarely encountered

REPRODUCTION
Unknown

DIET
Unknown. Presumed to be parasitoids of other insects

HELORID WASPS

This is a worldwide family with 1 extant genus, *Helorus*, containing 19 species, and 11 extinct genera.

Most species are known from the Palearctic region and adjacent areas of China. *Helorus* species prefer cooler environments, being absent from humid lowland tropical areas, although species have been described from the uplands of Sulawesi and Papua New Guinea.

Several species have been reared from larvae of green lacewings (Neuroptera: Chrysopidae) and presumably all species are specialized on these hosts.

Females of *Helorus paradoxus* live for about 5 weeks and over this period lay around 50 eggs individually into different host larvae. Wasp development takes about 30 days from egg-laying to emergence of the adult from the host cocoon.

BELOW | *Helorus anomalipes*, a species widespread across the northern hemisphere.

RIGHT | A female American *Pelecinus polyturator*, whose extraordinarily elongate metasoma is used to access the host beetle grubs living deep in soil.

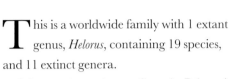

DISTRIBUTION
Worldwide, except for Arctic region, Indo-Pacific islands, and New Zealand.

GENUS
Helorus

HABITATS
Wide range of habitats, from forests to tundra

SIZE
1/8–3/16 in (3–5 mm) body length

ACTIVITY
Diurnal. Adults are usually found in only small numbers

REPRODUCTION
Females lay an egg into the lacewing larva within which the wasp larva hatches and consumes the host's body tissue after it has pupated

DIET
Solitary internal parasitoids of green lacewing larvae emerging from the host cocoon

PELECINID WASPS

This archaic family is restricted to the New World (North and South America) and comprises a single extant genus with three species. The family is much better represented in the Mesozoic fossil record by 18 extinct genera known from mostly Asian fossil formations.

These are large parasitoid wasps attacking dung beetle larvae, but due to their rarity are seldom seen.

The females have an exceptionally elongate metasoma, which is usually held in a characteristic curved position. The males have a shorter, club-shaped metasoma.

Pelecinus polyturator attacks Phyllophaga dung beetle larvae in North America, using its long metasoma to probe the soil to find the host beetle grub for egg-laying. The wasp larva develops internally but pupates outside the host. In North America males are rare and reproduction is mainly parthenogenetic.

DISTRIBUTION
New World (North and South America)

GENUS
Pelecinus

HABITATS
Forests

SIZE
³/₄–2 ³/₄ in (20–70 mm) body length

ACTIVITY
Diurnal. Adults occur at low density but are conspicuous in flight

REPRODUCTION
Females use their extremely long, slender abdomen to probe for and reach host larvae in the soil to lay an egg on

DIET
Parasitoids, attacking dung beetle larvae in the genus *Phyllophaga* (Coleoptera: Scarabaeidae)

PERADENIID WASPS

Peradeniids are an endemic Australian family comprising a single genus with two described living species (*Peradenia clavipes* and *P. micranepsia*). A fossil species is known from Eocene Baltic amber.

Species are rare and only known from a few specimens collected in the temperate forests of southeastern Australia and Tasmania.

Unusually for wasps, adults appear to be active only in winter.

BELOW | *Peradenia clavipes*, a rare species with a restricted distribution.

DISTRIBUTION
Australia, where they only occur in the forests of southeastern Australia and Tasmania

GENUS
Peradenia

HABITATS
Temperate forests

SIZE
$^3/_{16}$–$^3/_8$ in (5–10 mm) body length

ACTIVITY
Diurnal. Adults are rarely encountered

REPRODUCTION
Unknown

DIET
Unknown

PROCTORENYXID WASPS

This is a rare family of giant proctotrupoid wasps restricted to Asia, with only two genera and three species known (*Hsiufuropronia chaoi*, *Proctorenyxa incredibilis*, and *P. koreana*).

It was originally named Renyxidae, but the type genus name *Renyx* was preoccupied by a helminth (Cestoda) and the name required replacement.

ABOVE | *Proctorenyxa incredibilis*, a rare Asian species.

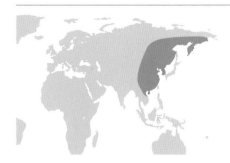

DISTRIBUTION
Asia (China, Russia, South Korea)

GENERA
Hsiufuropronia, Proctorenyxa

HABITATS
Forests

SIZE
$^1/_2$–$^9/_{16}$ in (12–15 mm) body length

ACTIVITY
Adults are rarely encountered

REPRODUCTION
Unknown

DIET
Parasitoid. Host unknown

PROCTOTRUPID WASPS

This globally distributed family has 33 genera and over 400 described species, and is most common in shaded or wet habitats.

Proctotrupid wasps are often parasitoids of beetle larvae living in leaf litter or rotten wood, with a few species attacking larvae of fungus gnats and one species even recorded as a parasitoid of a centipede, which is the only known example of a hymenopteran parasitoid of a myriapod and is in need of confirmation. The wasp larva pupates outside the host but remains attached by its hind end to the surface of the host and does not form a cocoon.

DISTRIBUTION
Worldwide

GENERA
Acanthoserphus, Afroserphus, Apoglypha, Austrocodrus, Austroserphus, Brachyserphus, Carinaserphus, Codrus, Cryptoserphus, Disogmus, Exallonyx, Fustiserphus, Glyptoserphus, Heloriserphus, Hemilexodes, Hormoserphus, Maaserphus, Mischoserphus, Nothoserphus, Oxyserphus, Paracodrus, Parepyris, Parthenocodrus, Phaenoserphus, Phaneroserphus, Phoxoserphus, Proctotrupes, Pschornia, Serphonostus, Smithoserphus, Trachyserphus, Tretoserphus, Tropidopria

HABITATS
A wide range of habitats but most abundant in woodlands

SIZE
$1/8$–$9/16$ in (3–15 mm) body length

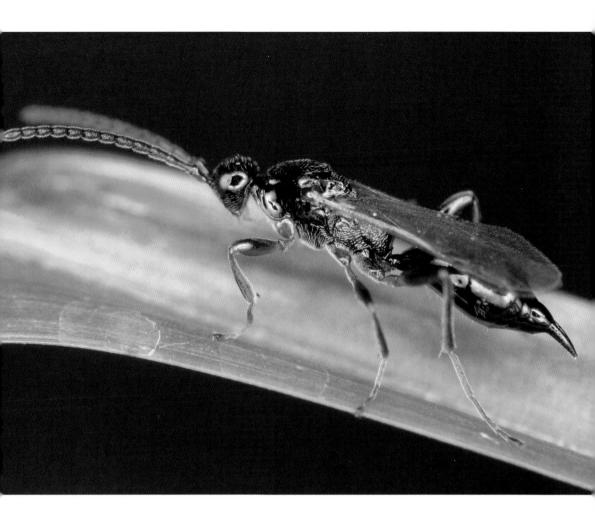

ABOVE | *Exallonyx alticola,*
an African species whose host
is unknown.

LEFT | *Brachyserphus parvulus,*
a species widely distributed across
the northern hemisphere attacking
hosts in several beetle families.

ACTIVITY
Mostly diurnal but some species
regularly come to lights at night. Adults
can be reasonably numerous

REPRODUCTION
Females use their moderately elongate,
downcurved last abdominal segments
to probe for hosts for egg-laying

DIET
Parasitoids attacking beetle larvae or
fungus gnats (Diptera: Mycetophilidae),
sometimes gregariously

Proctotrupidae are all endoparasitoids (internal)
and can be gregarious, such as *Codrus carolinensis*
attacking larvae of the rove beetle *Platydracus violaceus*
(Staphylinidae), or solitary, as in *Exallonyx
philonthiphagus* attacking the smaller mature larvae of
the rove beetle *Philonthus turbidus.* Another gregarious
parasitoid, *Phaenoserphus viator,* attacks carabid
ground beetles and survives winter in the egg stage.
Adult proctotrupid wasps emerge from their pupa
after one to two weeks.

ROPRONIID WASPS

Roproniid wasps are a rare family with 2 living genera containing 24 species distributed over Asia, southeast Europe (Turkey), and North America, and 7 extinct genera known from fossils dating back to the Jurassic, Cretaceous, and Paleocene epochs.

The genus *Xiphyropronia* is only known from China, while *Ropronia* species are found more widely.

Ropronia brevicornis attacks the sawfly *Periclista*, emerging from the pupa.

BELOW | *Ropronia garmani*, a North American species occurring in forests near streams.

DISTRIBUTION
Northern hemisphere (eastern Palearctic and North America)

GENERA
Ropronia, Xiphyropronia

HABITATS
Tropical or temperate forests

SIZE
3/16–1/2 in (5–12 mm) body length

ACTIVITY
Diurnal. Adults are rarely encountered

REPRODUCTION
Biology is poorly known but thought to be endoparasitoids ovipositing in larvae and emerging from the pupa

DIET
Parasitoids of sawflies

VANHORNIID WASPS

This rare family has a single genus containing four described species and is found in Europe, North America, and Central and East Asia, south as far as Thailand. Undescribed species are also known from Japan and the USA.

These are unusual wasps in that they have exodont mandibles (teeth project outwards) that may have evolved to enable the adult wasp to cut itself out of its host pupa or to chew through wood to reach its host for egg-laying, given that most of the false click beetle (Eucnemidae) hosts are associated with decaying wood.

Vanhornia eucnemidarum attacks false click beetle larvae developing in old, dying maple (*Acer*) trees. When exserted, the long ovipositor is held pointing forward in a groove running along the middle of the underside of the metasoma. The host beetles are rare, explaining why this family of wasps is also rare.

BELOW | A female *Vanhornia* ovipositing into a host click beetle larva feeding inside a decaying tree trunk.

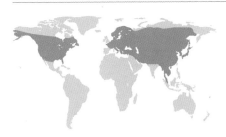

DISTRIBUTION
Northern hemisphere (North America, Europe, Asia)

GENUS
Vanhornia

HABITATS
Tropical or temperate forests, woodland, scrub

SIZE
$^3/_{16}$–$^1/_4$ in (4–7 mm) body length

ACTIVITY
Diurnal. Adults are rarely encountered although have been noted

REPRODUCTION
Unknown, but females will need to access their host beetle larvae that are feeding in rotten wood

DIET
Parasitoids of false click beetle larvae (Eucnemidae)

SHELF-FACED WASPS

This diverse and commonly collected family of parasitoid wasps contains 190 genera and 2,048 described species in 3 subfamilies: Ambositrinae, Belytinae, and Diapriinae. Although Ambositrinae are mostly found in the southern hemisphere, the other subfamilies are found worldwide.

These wasps are usually easily identified by a shelf-like projection on the front of the face from which the antennae arise, in combination with distinctive wing venation, or near lack of venation (subfamily Diapriinae).

Sexual dimorphism can be extreme in shelf-faced wasps, especially in some species with flightless females. Species of several genera can be wingless or have reduced wings, an adaptation to their lifestyle associated with the ground-level environment, including seashores in the case of *Platymischus*.

Some groups of diapriids specialize on cyclorrhaphan Diptera hosts, others (particularly

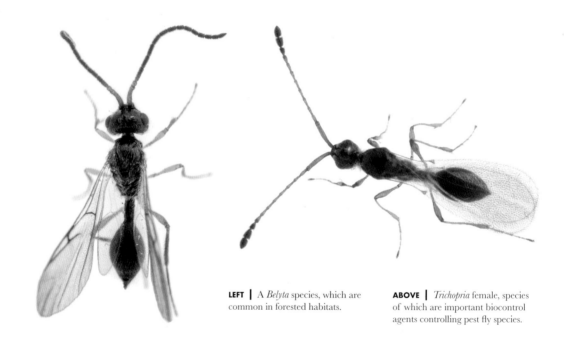

LEFT | A *Belyta* species, which are common in forested habitats.

ABOVE | *Trichopria* female, species of which are important biocontrol agents controlling pest fly species.

DISTRIBUTION
Worldwide

GENERA
190 genera in 3 subfamilies: Ambositrinae, Belytinae, Diapriinae. Some of the common genera include *Ambositra, Aclista, Basalys, Belyta, Coptera, Entomacris, Spilomicrus, Trichopria*

HABITATS
A range of habitats, but more common in leaf litter associated with forests or damp habitats

SIZE
$1/32$–$5/16$ in (1–8 mm) body length

ACTIVITY
Adults are common in the leaf litter, understory, or in the soil surface environment

Belytinae and Ambositrinae) on nematocerous Diptera hosts, such as fungus gnats. There is at least one record of a diapriid parasitizing a rove beetle (Coleoptera: Staphylinidae), and a few species can act as hyperparasitoids of dipteran or hymenopteran parasitoids that attack flies.

Some species are specialized termite associates, developing within termitaria, while others are myrmecophiles, being associated with ants. Others utilize hosts in aquatic habitats. Most species are usually found where there is plenty of decaying vegetation and fungi that their host flies are associated with.

Some diapriids are important biocontrol agents, with many species attacking pest species of insects. *Trichopria painteri* attacks the Stable Fly (*Stomoxys calcitrans*, Muscidae), a blood-sucking filth fly pest of mammals; *T. anastrephae* attacks the Spotted Wing Drosophila fruit fly (*Drosophila suzukii*), an economic pest of thin-skinned fruit crops; and *T. drosophilae* attacks a range of fruit fly species. An African species, *Coptera silvestrii*, was introduced to control Mediterranean Fruit Fly (*Ceratitis capitata*) in Europe and Hawaii. Others have a negative impact, such as *Trichopria columbiana*, which attacks aquatic fly species (*Hydrellia*) that control invasive waterweeds (*Hydrilla* species).

REPRODUCTION
Often solitary, although many species of Diapriinae are gregarious, females laying many eggs in the host larva or pupae. Mating occurs close to the host soon after emergence of the adults, sometimes with a couple of hundred individuals produced from a single host

DIET
Internal parasitoids of the larvae and pupae of mostly flies (Diptera)

ISMARID WASPS

T his globally widespread family was until recently considered to be a subfamily within the Diapriidae, containing the single genus *Ismarus*, with about 60 species.

Ismarids lack the facial shelf of the Diapriidae.

These wasps are hyperparasitoids, attacking larvae of Dryinidae that are developing on leafhoppers (Cicadellidae). Given that they attack dryinids, which are useful biocontrol agents of pest planthoppers, the ismarids may have a detrimental effect on their efficiency.

Ismarid species never seem to be particularly common, occurring in woodland or forests at higher elevations in tropical areas and at lower elevations in temperate regions.

At least one species, *Ismarus halidayi*, is common and widespread, having been recorded from Asia, Europe, North America, and South Africa.

BELOW | *Ismarus* species are hyperparasitoids of leafhoppers (Cicadellidae) via Dryinidae.

DISTRIBUTION
Worldwide

GENUS
Ismarus

HABITATS
Forests, woodland

SIZE
$^1/_{32}$–$^1/_8$ in (1–3 mm) body length

ACTIVITY
Seem to be diurnal; infrequently encountered

REPRODUCTION
Solitary internal (endo)parasitoids

DIET
Hyperparasitoids feeding on larva of Dryinidae, which are parasitoids of leafhoppers (Cicadellidae)

TRICKSTER WASPS

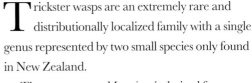

Trickster wasps are an extremely rare and distributionally localized family with a single genus represented by two small species only found in New Zealand.

The genus name *Maaminga* is derived from a Māori word meaning "a trickster" or "mystifying" because the wasps evaded discovery for so long. These wasps occur on the forest floor, in low scrub and tussock grasses.

Maaminga rangi is a slight, fully winged species that is common in the North Island forests.

Maaminga marrisi is a stockier species with individuals that may be fully winged or brachypterous (reduced wings), and is found in coastal environments including islands, in scrub or forests, or in grass tussocks in alpine environments above the snow line.

Maaminga marrisi was named after John Marris, who collected this species during a survey of rare beetles on an island in Cook Strait. *Maaminga rangi* was named after Ranginui, the sky father of the Māori creation story.

BELOW | The New Zealand Trickster Wasp, *Maaminga marrisi*.

DISTRIBUTION
New Zealand

GENUS
Maaminga

HABITATS
Forests, coastal scrub, alpine snow tussocks

SIZE
$^1/_{32}$–$^1/_{16}$ in (1–2 mm) body length

ACTIVITY
Adults can be common in forests; one species has shortened wings and is probably restricted to the leaf litter habitat

REPRODUCTION
Unknown

DIET
Unknown

MONOMACHID WASPS

Monomachids are represented by 2 genera and 30 species.

This is a family of rare species with localized distributions, occurring only in the southern hemisphere in Australia, New Guinea, and South America. This contemporary distribution suggests they are a relict family with a previous Gondwanan distribution and historically were probably more diverse, although not many fossils are known.

The species are rather bizarre, having massive mandibles, and are sexually dimorphic—females have a thin, elongate, sickle-shaped metasoma with a short ovipositor hidden within the metasoma, and males have an elongate, petiolate club-shaped metasoma.

Some of the New World species are light green, an unusual color in wasps, sometimes seen in unrelated wasp families in Chile and Madagascar.

DISTRIBUTION
Australia, New Guinea, Neotropical region from Mexico to Chile

GENERA
Chasca, Monomachus

HABITATS
Tropical or temperate forests, woodland, scrub

SIZE
3/8–3/4 in (10–20 mm) body length

ACTIVITY
Adults can be common and are mostly active in winter

REPRODUCTION
Females lay an egg directly into the host fly egg; the wasp larva develops in the host larva and emerges from the larva or pupa

DIET
Parasitoid of eggs, larvae, and pupae
of fly species in the genera *Boreoides*
and *Chiromyza* (Diptera:
Stratiomyidae)

A couple of species have reduced wings in the
females. They are associated with cooler, wet forest
habitats, usually at higher elevation.

These wasps are parasitoids of soldier flies
(Stratiomyidae). *Monomachus antipodalis* is an
egg-larval or egg-pupal parasitoid of species
of *Boreoides*.

Monomachids have potential as biocontrol
agents, with the New World species *M. fuscator* and
M. eurycephalus known to attack *Chiromyza vittata*,
the larvae of which are pests feeding on roots of
coffee plants in Brazil.

GEOSCELIONID WASPS

These wasps are an ancient family containing two rare, living genera represented by three species, and three extinct genera with five species known from the Cretaceous and Eocene.

Plaumannion fritzi is rare, known only from three specimens collected from leaf litter in the Amazon rain forest of Brazil. *Plaumannion yepezi* is only known from a single female collected in a yellow pan trap in a Venezuelan cacao plantation. *Huddlestonium exu* is an African species known from a handful of specimens collected in grassland in Ivory Coast, Kenya, and São Tomé.

The biology of geoscelionid wasps is unknown. This is a newly established family, described in 2021 based on a thorough phylogenetic appraisal of the platygastroid wasps, and was elevated from tribal rank in the Scelionidae.

LEFT | The African *Huddlestonium exu.*

DISTRIBUTION
Africa: Kenya, Ivory Coast, São Tomé;
South America: southeastern Brazil,
Venezuela

GENERA
Huddlestonium, Plaumannion

HABITATS
Tropical forests, woodland, grassland

SIZE
$1/16$ in (2 mm) body length

ACTIVITY
Adults are rarely encountered

REPRODUCTION
Unknown

DIET
Unknown

JANZENELLID WASPS

A rare and unusually restricted family only known so far from Costa Rica, janzenellid wasps are represented by a single genus and one living species, *Janzenella innupta*, which is also known from the fossil record in Dominican amber (Miocene). A second Eocene fossil species, *J. theia*, is known from Baltic amber.

The biology of this family is unknown and males have not been collected.

Janzenella innupta is an unusual-looking platygastroid wasp, being tiny and with a strongly elongate and compressed body, superficially resembling a flat wasp (Bethylidae).

The family Janzenellidae was only established in 2021 based on a thorough phylogenetic appraisal of the platygastroid wasps.

BELOW | *Janzenella theia*, a fossil species embedded in Baltic amber.

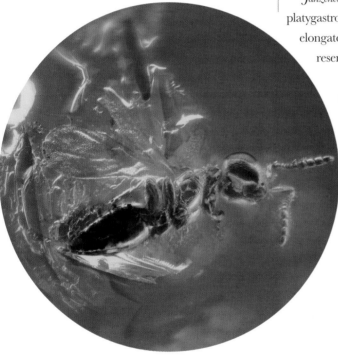

DISTRIBUTION
Costa Rica

GENUS
Janzenella

HABITATS
Dry deciduous forests, or disturbed areas

SIZE
$^1/_{32}$ in (0.8 mm) body length

ACTIVITY
Unknown. Adults are rarely encountered

REPRODUCTION
Unknown

DIET
Unknown, but likely to be parasitoids of insect eggs

NEUROSCELIONID WASPS

This is another range-restricted family known only from a single genus with seven described species occurring in southeastern Asia and eastern Australia. Two fossil genera are known.

For many years the genus *Neuroscelio*, described in 1913, was only known by the single female holotype specimen of *N. nervalis* that was swept in a North Queensland forest. Subsequent collecting in the 1990s revealed the genus to be common in Australia with an additional four species named. *Neuroscelio* was recorded from the Oriental region with the description of two further species in 2009.

This is another platygastroid family established in 2021 based on a thorough phylogenetic appraisal of platygastroid wasps. The genus was previously placed in the Scelionidae but was elevated to family rank after this assessment of evolutionary relationships.

ABOVE | The Australian
Neuroscelio doddi.

DISTRIBUTION
Australia, Vietnam, Thailand, Malaysia
(Sarawak)

GENUS
Neuroscelio

HABITATS
Tropical or temperate forests,
woodland

SIZE
³/₁₆ in (4 mm) body length

ACTIVITY
Unknown

REPRODUCTION
Unknown

DIET
Unknown

NIXONIID WASPS

Nixoniid wasps are a rarely collected family represented by a single genus with 16 species only known from Africa and the Oriental region.

This family includes some of the largest platygastroid wasps, with a couple of species reaching $^3/_8$ in (9 mm) in length. Adults are active in the ecologically inhospitable dry summer season in southwestern Africa.

The single host record suggests they are parasitoids of eggs of Armored Ground Crickets, one species having been reared from eggs of *Acanthoplus discoidalis* (Orthoptera: Tettigoniidae, Hetrodinae).

The genus was previously placed in the Scelionidae but was elevated to family rank based on a thorough phylogenetic appraisal published in 2021.

ABOVE | A male *Nixonia masneri* from South Africa. *Nixonia* are parasitoids of Armored Ground Crickets (*Acanthoplus discoidalis*).

DISTRIBUTION
Africa, Southeast Asia: India, Laos, Sri Lanka, Thailand, Vietnam

GENUS
Nixonia

HABITATS
Tropical or temperate forests, woodland, dry scrub, fynbos

SIZE
$^3/_{16}$–$^3/_8$ in (4–9 mm) body length

ACTIVITY
Adults are rarely encountered

REPRODUCTION
Females lay their eggs into the eggs of the host crickets that are deposited in soil

DIET
Parasitoid of eggs of Armored Ground Crickets (*Acanthoplus discoidalis*)

PLATYGASTRID WASPS

Platygastrid wasps are a commonly collected and diverse, worldwide family of parasitoid wasps containing 60 genera and more than 4,000 species in 2 subfamilies: Platygastrinae and Sceliotrachelinae.

Platygastrinae are mostly parasitoids of gall midge (Cecidomyiidae) eggs, developing only after the host has reached pupal or prepupal stage.

Sceliotrachelinae are parasitoids of beetle (Coleoptera) or bug (Homoptera) eggs, or bug nymphs, or crabronid wasp larvae.

Species can be common and are generally very small, averaging around $^{1}/_{32}$ in (1 mm) in length, a size circumscribed by the small food resource available for their developing larvae. The genus *Inostemma* has a characteristic, long, curved,

DISTRIBUTION
Worldwide

GENERA
60 genera, including *Orwellium*.
Some of the more common include *Allotropa, Inostemma, Leptacis, Platygaster, Sceliotrachelus, Synopeus*

HABITATS
Wide range of habitats, from forests to tundra

SIZE
$^{1}/_{64}$–$^{1}/_{16}$ in (0.5–2 mm) body length

ACTIVITY
Adults are common

REPRODUCTION
Females lay eggs into the eggs of the host and usually only develop after the host larva has reached late larval or pupal stage. One species (*Platygaster zosine*) is polyembryonic, with one egg dividing into multiple embryos.

Larvae of some platygastrids are
among the most bizarre of wasp
larvae, with huge mandibles and two
long "tail" projections

DIET
Parasitoids of insect eggs and larvae,
feeding on the body contents of their
host

horn-like structure projecting from the top of the
metasoma forward over the rest of the body. This is
an evolutionary adaptation to internally house a
very long ovipositor.

Several species are economically important,
being used as biocontrol agents of pests. Many
of these attack hemipteran pests of citrus. *Amitus*
species attack whiteflies (Aleyrodidae), and *Allotropa*
species attack mealybugs (Pseudococcidae) and
cochineals (Dactylopiidae). An Australian species,
Aphanomerus pusillus, controls the Pandanus
Planthopper (*Jamella australiae*), which ravages the
Screwpine (*Pandanus tectorius*) in coastal regions.

SCELIONID WASPS

OPPOSITE | A scelionid attacking eggs of a shield bug. Her larvae will develop within the egg, devouring the contents.

BELOW | A *Telenomus* female ovipositing into a host insect egg.

T his extremely diverse and species-rich family comprises about 150 extant genera and about 30 fossil genera, with more than 3,000 species contained in 3 subfamilies: Scelioninae, Teleasinae, and Telenominae, distributed worldwide.

Many species, particularly in the genera *Gryon*, *Telenomus*, and *Trissolcus*, are economically important, being used as biocontrol agents of pest insects. Several species in these three genera attack the eggs of the Painted or Bagrada Bug

DISTRIBUTION
Worldwide

GENERA
150 genera in 3 subfamilies:
Scelioninae, Teleasinae, Telenominae

HABITATS
Wide range, from forests to tundra

SIZE
$^1/_{64}$–$^3/_8$ in (0.5–10 mm) body length

ACTIVITY
Diurnal. Adults are common

REPRODUCTION
Females lay a single or several eggs directly into the host egg with their hypodermic ovipositor. The wasp larva pupates within the host egg shell

DIET
Internal parasitoids of insect and spider eggs, the wasp larvae feeding on the egg contents

The tiny ¹/₃₂ in (1 mm) South African wingless species *Echthrodesis lamorali* is highly specialized, being only found in the intertidal zone on rocky shores, where it attacks eggs of spiders living within old shells in the underwater marine environment. Two other genera attack aquatic insects—*Tiphodytes* and *Thoron* parasitize the eggs of the water bugs (Gerridae and Nepidae).

A good number of species are leaf litter inhabitants and these exhibit morphological adaptation to this environment, often being wingless and compact with fused body segments, such as *Baeus*, *Encyrtoscelio*, *Parabaeus*, and *Platyscelidris*, and hence do not look like typical wasps at all.

Species in several genera such as *Mantibaria*, *Protelenomus*, *Sceliocerda*, and *Telenomus* exhibit phoresy, where the females hitch a ride and attach themselves by biting into the abdominal intersegmental membrane on the adult female host insect (commonly grasshoppers, mantids, or bugs) until she lays her eggs, which are then attacked.

(*Bagrada hilaris*, Pentatomidae), a serious pest of Brassicaceae crops. *Trissolcus basalis* successfully controls the Southern Green Stinkbug (*Nezara viridula*), which feeds on a wide variety of grain, vegetable, fruit, cotton, and nut crops.

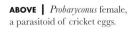

ABOVE | *Probaryconus* female, a parasitoid of cricket eggs.

SPARASIONID WASPS

Sparasionid wasps are an ancient group of wasps found worldwide with 5 genera and 169 species, plus 1 fossil genus (*Electroteleia*) known from Burmese amber (Cretaceous).

Archaeoteleia is likely a Gondwanan relict genus occurring in Chile and New Zealand. *Mexon, Listron*, and *Sceliomorpha* are only known from the Americas; *Sparasion* is more widely distributed, with the exception of Australia and South America.

Sparasionidae are large wasps, some reaching a similar size—up to ½ in (12 mm)—to *Nixonia* (Nixoniidae) or *Triteleia* (Scelionidae) species. Based on the few known host records, sparasionids are egg parasitoids of long-horned grasshoppers (Tettigoniidae) or crickets (Rhaphidophoridae). *Sparasion* species are known to attack eggs of Tettigoniidae in North America, and *Archaeoteleia* species attack eggs of the wetas (giant flightless crickets) in New Zealand. The relatively large size of the host insect eggs provides a more abundant food resource for the developing wasp larvae that allows these wasps to reach above-average body size for Platygastroidea.

Although most platygastroid wasps are rather uniformly black or dark brown, several *Sparasion* species have brilliant metallic coloration. Some of the *Sceliomorpha* species are brightly colored with red and yellow.

LEFT | *Archaeoteleia gilbertae*, a species endemic to New Zealand. *Archaeoteleia* is a Gondwana relict genus, restricted to Chile and New Zealand, with species occurring nowhere else in the world.

RIGHT | Frontal view of the head of a *Sparasion* species, which is a parasitoid of long-horned grasshopper eggs.

DISTRIBUTION
Worldwide

GENERA
Archaeoteleia, Mexon, Listron, Sceliomorpha, Sparasion

HABITATS
Wide range, from forests to tundra

SIZE
³/₁₆–¹/₂ in (5–12 mm) body length

ACTIVITY
Adults are rarely encountered

REPRODUCTION
Females lay their eggs directly into the host insect egg

DIET
Parasitoids, larvae feeding on the contents of grasshopper and cricket eggs

AUSTRALIAN GALL WASPS

The family Austrocynipidae is restricted to Queensland in Australia and contains a single species: *Austrocynips mirabilis*. Possibly the rarest family of wasps, only three specimens have ever been collected.

From an evolutionary perspective this family is sister to all the other Cynipoidea, which share a common ancestor, meaning it is an ancient lineage.

The only known specimens were reared from cones of Hoop Pine (*Araucaria cunninghamii*) that were infested with oecophorid moths, the distributional range of which extends along the coastal region from northern Queensland to northern New South Wales. This pine species occurs in coastal tropical and subtropical rain forests over an elevation range from sea level to 3,300 ft (1,000 m). The Hoop Pine was commercialized and historically cones were collected for propagation and timber production, but this is no longer done. The rarity of the species is likely to be related to how difficult these cones are to access in these extremely tall trees (130–195 ft [40–60 m] high).

Other species of *Araucaria* were surveyed at the time of discovery of this enigmatic species, but no related species of *Austrocynips* were found. It is likely to be a parasitoid species that may attack oecophorid moth caterpillars developing in Hoop Pine cones. There is a possibility that the species may be phytophagous, forming galls or acting as a seed predator within the cones.

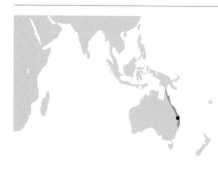

LEFT | The endemic Australian family Austrocynipidae contains a single species, *Austrocynips mirabilis*, which probably attacks moth caterpillars developing in the cones of Hoop Pine (*Araucaria cunninghamii*).

DISTRIBUTION
Only known from a single locality in Queensland, Australia (black dot); however, the distribution of *Austrocynips mirabilis* could mirror the distribution of the host Hoop Pine, the range of which is illustrated here

GENUS
Austrocynips

HABITATS
Coastal tropical and subtropical rain forest

SIZE
3/16 in (3.5 mm) body length

ACTIVITY
Unknown

REPRODUCTION
Unknown

DIET
Unknown

GALL WASPS

T he gall wasps are a group of about 1,200 described species classified in 12 tribes and 83 genera, but with many species still to describe.

Species richness is centered in the northern hemisphere, with only a few representatives known from the southern hemisphere.

Cynipidae either form galls on a variety of plants, such as the large radiation of oak gall wasps, or develop in galls formed by other wasps, or more rarely in galls formed by midges or moths. The gall wasps induce a huge variety of galls, often structurally complex with elaborate external architecture, on oaks, roses, grasses, and herbaceous plants. The ovipositing female produces cellulases that lyse plant tissues, presumably to help the larva digest woody tissue, but it is not yet known how the gall is induced and whether this is by the adult female's venom or the wasp larva.

ABOVE RIGHT | The California Gall Wasp (*Andricus quercuscalifornicus*) a gall inducer on species of *Quercus* (the White Oaks).

RIGHT | The Marble Gall Wasp (*Andricus kollari*) forms large, round marble-like galls on Oak trees (*Quercus* spp.).

DISTRIBUTION
The family is most prevalent in the northern hemisphere; also present in South America and Africa. New Zealand has an introduced species

GENERA
83 genera in 12 tribes

HABITATS
Mostly temperate forests, woodland, fynbos, and grassland, but a few species are more tropical, occurring in savanna ecosystems

SIZE
On average $^1/_{32}$–$^1/_{16}$ in (1–2 mm) long

ACTIVITY
Diurnal or nocturnal as adult wasps

REPRODUCTION
Cynipidae develop in galls

DIET
Gall wasp larvae feed on endosperm tissue in the host plant galls. Adults probably only live for a few days and may imbibe nectar

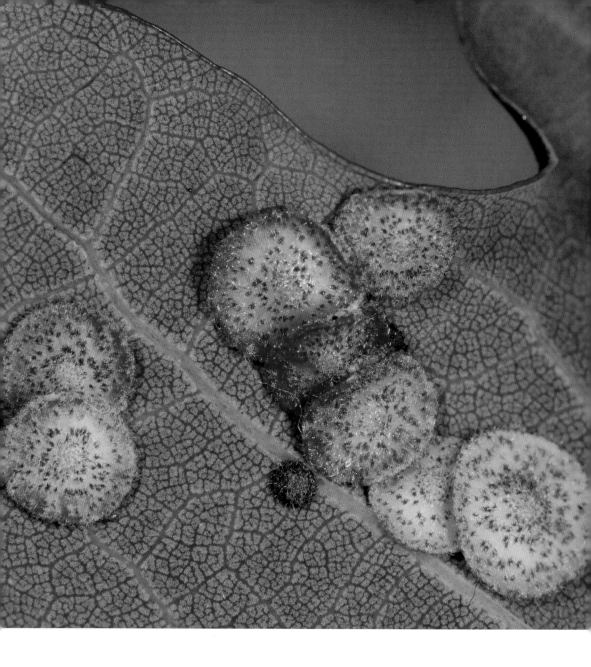

Many form conspicuous galls on oaks and are accordingly well known, with English names such as the Spangle Gall (*Neuroterus quercusbaccarum*) and "oak apples," referring to a variety of species such as the California Gall Wasp (*Andricus quercuscalifornicus*) or, in Europe, *Biorhiza pallida*.

The South African endemic genus *Rhoophilus* forms hard woody galls on *Searsia*, modifying a primary gall instigated by a moth, and in the process killing the moth caterpillar.

ABOVE | Spangle galls of *Neuroterus quercusbaccarum* on the underside of oak leaves represent the over-wintering asexual generation; the sexual generation produces smooth current galls formed on the oak catkins in spring.

A few gall wasps are used in the biological control of invasive plants. The Hieracium Gall Wasp (*Aulacidea subterminalis*) has been used to control invasive weeds in both North America and New Zealand.

Other species, such as the Chestnut Gall Wasp (*Dryocosmus kuriphilus*), are invasive pests themselves. This species, indigenous to China, has become a serious horticultural and agricultural pest species of Sweet Chestnut (*Castanea sativa*) in many other areas of the northern hemisphere.

Many oak gall wasps have an alternation of generations, with both sexes produced in one generation and then only females (thelytoky) in the next. Different generations produce different gall structures on different parts of the tree.

Galls are distinctive to each gall wasp but the mechanism by which the gall wasp induces a particular pattern of cell growth in its host plant is currently unknown. The Magna Carta, among many other texts, was written in iron gall ink, which for centuries has been produced using tannins from oak galls.

ABOVE | The Chestnut Gall Wasp (*Dryocosmus kuriphilus*), a species indigenous to China, has become a serious horticultural and agricultural pest species.

RIGHT | *Dryocosmus kuriphilus* forms deleterious galls on Sweet Chestnut (*Castanea sativa*), resulting in fruit production losses of up to 80 percent.

FIGITID WASPS

The family Figitidae has a global distribution with the exception of the Arctic and Antarctic regions. It is a species-rich family with over 1,700 described species in the world, and many more still requiring description, with 157 genera placed in 12 subfamilies.

The Figitidae are parasitoid wasps that attack a wide range of insect hosts, either as primary parasitoids, as hyperparasitoids, or as inquilines inside galls. Species of Aspicerinae, Eucoilinae, and Figitinae attack fly larvae developing in various niches, such as leaf-miners, or in seaweed, dung,

ABOVE | A parasitoid eucoiline wasp that attacks fly larvae or pupae, thereby controlling pest species.

RIGHT | Males of *Leptopilina japonica* aggregating on a flower cluster to obtain nectar for their energy requirements.

DISTRIBUTION
Worldwide with the exception of the Arctic and Antarctic regions

GENERA
157 genera in the subfamilies Anacharitinae, Aspicerinae, Charipinae, Emargininae, Euceroptrinae, Eucoilinae, Figitinae, Mikeiinae, Parnipinae, Plectocynipinae, Pycnostigminae, Thrasorinae

HABITATS
Wide range, from tropical forests and savanna ecosystems to cold or arid temperate environments

SIZE
On average $1/32$–$1/16$ in (1–2 mm) in length

ACTIVITY
Diurnal or nocturnal as adult wasps,
larvae develop within host insect
larvae

REPRODUCTION
Females lay eggs into host insect
larvae, which may be free-living,
gall-formers, or parasitoids of bugs.
Larval development is internal
(endoparasitoid), with emergence from
the cocooned final instar or pupal host

DIET
Figitid wasp larvae feed on body
tissue of the host insect larva they are
developing in. Adult wasps probably
live for only a few days, and imbibe
nectar or host body fluids for their
energy requirements

APOCRITA (PARASITICA)—Parasitoid Wasps

or carrion. The Anacharitinae attack lacewing larvae (Neuroptera); the Parnipinae and Thrasorinae are parasitoids of cynipoid or chalcidoid gall-inducing wasps. The Charipinae are hyperparasitoids of aphids and psyllids (Hemiptera) through other parasitoid wasps (Braconidae or Chalcidoidea). Nothing is known about the biology of the Emargininae or Pycnostigminae.

Many of the host insect larvae are pest species and the wasps perform a useful biocontrol agent role of agricultural and forestry industry pests. *Aganaspis daci* is a successful biocontrol agent of the Caribbean Fruit Fly (*Anastrepha suspensa*) and Mediterranean Fruit Fly (*Ceratitis capitata*) (Tephritidae). The Asian species *Leptopilina japonica* has been introduced in many countries to control the Spotted Wing Drosophila fruit fly (*Drosophila suzukii*), a pest of soft fruits worldwide. *Trybliographa rapae* controls the Cabbage Root Fly (*Delia radicum*).

The South African *Pycnostigmus mastersonae* is the only known metallic species of Cynipoidea; most species are of sober color, although morphologically they are varied enough that the subfamilies Charipinae and Eucoilinae have been regarded as separate families in the past.

ABOVE | A female *Melanips* species attacking a hoverfly larva (Syrphidae). The wasp larva that hatches from the egg she is laying will consume the body contents, killing the fly larva. This is a negative impact by the wasp as hoverfly larvae are beneficial as they feed on pest insects such as aphids.

LEFT | A female *Leptopilina japonica* laying eggs inside a raspberry to parasitize Spotted Wing Drosophila fruit fly (*Drosophila suzukii*) maggots that infest and ruin fruit. This is an example of classical biological control, implemented through the introduction of this indigenous Asian wasp species to other parts of the world, thereby reducing the need for insecticide use.

IBALIID PARASITOID WASPS

T his family has an indigenous northern hemisphere distribution, but has been introduced to Australia, New Zealand, and South Africa. Ibaliidae include 21 species in 3 genera, in 2 subfamilies (Eileenellinae and Ibaliinae), with highest species richness in North America. The Eileenellinae subfamily is only represented by a single species in Papua New Guinea, *Eileenella catherinae*, which has not been collected again since the species was described in 1992.

RIGHT | *Ibalia anceps*, which attacks horntail sawflies in the genus *Tremex*.

BELOW | *Ibalia leucospoides*, a biocontrol agent of the Sirex Woodwasp (*Sirex noctilio*), a pest in pine plantations.

DISTRIBUTION
A northern hemisphere family introduced to Australia, New Zealand, and South Africa

GENERA
Eileenella, Heteribalia, Ibalia

HABITATS
Temperate (more rarely tropical) forests, woodland, deserts

SIZE
On average $^3/_{16}$–$^3/_8$ in (5–10 mm) in length

ACTIVITY
Diurnal or nocturnal as adult wasps, larvae develop in woodwasp larvae boring inside wood

REPRODUCTION
Parasitoids of wood-boring woodwasp larvae. Males mate with females while they are laying eggs

Ibaliinae are parasitoids of wood-boring woodwasp larvae (Anaxyelidae, Siricidae). The female inserts her ovipositor down the tunnel bored in pine trees by the host woodwasp larva, to lay an egg either into the egg of the host or into the young host larva. On hatching, the ibaliid wasp larva feeds internally, as an endoparasitoid, until the third instar, when it emerges and feeds externally on the host larva. These are large cynipoid wasps as their host larvae are relatively large compared to hosts of the other cynipoid parasitoids.

Ibaliids are beneficial parasitoid wasps that attack pest insects in the forestry industry. The Sirex Ibaliid Wasp (*Ibalia leucospoides*, Ibaliidae) controls the invasive Sirex Woodwasp (*Sirex noctilio*) (Siricidae), a pest in Monterey Pine plantations (*Pinus radiata*).

DIET

Wasp larvae feed on body tissue of the host woodwasp larva they are developing in. Adult wasps are short-lived and imbibe nectar for their energy requirements

LIOPTERID PARASITOID WASPS

OPPOSITE | The endemic Peruvian genus *Peras*.

BELOW | *Paramblynotus*, a species-rich widespread genus, whose biology is still unknown.

This archaic family has a worldwide distribution except for the northern Palearctic region (Europe and northern Asia) and northern parts of North America. Species richness is concentrated in the tropics. This group of wasps is known from 143 species in 10 genera, of which *Paramblynotus* contains by far the most species.

Liopterid parasitoid wasps are a rarely encountered family, with many species only known from single specimens. Species of *Paramblynotus*, however, may be fairly abundant at certain times of the year.

DISTRIBUTION
Worldwide, with the exception of the northern parts of the Holarctic and Palearctic regions, the southern tip of South America, Tasmania, and New Zealand

GENERA
Dallatorrella, Kiefferiella, Liopteron, Mesocynips, Oberthuerella, Paramblynotus, Peras, Pseudibalia, Tessmannella, Xenocynips

HABITATS
Wide range, from tropical forests and savanna ecosystems to cold, arid, and temperate environments

SIZE
On average $1/16$–$1/8$ in (2–3 mm) in length, some reaching $3/8$ in (10 mm)

ACTIVITY
Diurnal or nocturnal as adult wasps, larvae probably endoparasitoids within host insect larvae

The biology of the family is unconfirmed, but they are probably parasitoids of wood-boring insects (Coleoptera: Buprestidae; Hymenoptera: Siricidae), as they have been collected from logs or trees infested with xylophagous (wood-feeding) insect larvae, and hence may have potential as biocontrol agents of pests of the forestry industry.

Two species of *Kiefferiella* were reared from buprestid larvae (*Acmaeodera pulchella*) developing in logs, *Paramblynotus yangambicolous* was reared from a rotten log of Euphorbiaceae, and *Kiefferiella* and *Paramblynotus* species were reared from trees in the family Fabaceae.

Species are strong-bodied with thick exoskeletons, and many have strong, backward-projecting ridges on the body, suggesting that these help with negotiating confined substrates such as rotten logs when entering wood or emerging from their pupation sites.

Paramblynotus species may have a modified area around the spiracle to trap air, enabling the wasp to continue with respiration when in a confined and possibly wet substrate. Only the females have these excavations, suggesting that it is a modification related to searching for hosts for egg-laying, rather than a requirement for exiting as adults from concealed hosts that their larvae have developed within.

REPRODUCTION
Biology is unconfirmed, but females probably lay eggs into concealed wood-boring host larvae

DIET
Liopterid wasp larvae will probably feed on body tissue of the host insect larva they are developing in. As with other parasitoid wasps, adults probably imbibe nectar for their energy requirements

FIG WASP POLLINATORS

This family has a pan-tropical distribution that extends into temperate areas. The *true* fig wasp pollinators in the subfamilies Agaoninae, Kradibiinae, and Tetrapusiinae include 19 genera and 358 described species, although the majority of species are still undescribed. The subfamily Sycophaginae was recently formally transferred from Agaonidae to Pteromalidae.

Agaonid fig wasps are the sole pollinators of fig trees in the genus *Ficus* (Moraceae), of which there are more than 750 species in the world, but many fig wasp pollinators are still unknown.

Females enter receptive figs through the ostiole, a narrow bract-lined opening at the apex of the fig, to lay eggs down the floret styles into the ovary, in the process pollinating the florets lining the inside of the cavity. Fig wasps can breed nowhere else, except for inside figs, and figs cannot reproduce without these specialist fig wasp pollinators, a relationship that is a classic example of an obligate mutualism that has evolved over the last 80 or so million years. The species-level relationship between the two partners is generally tight, and often considered to be a classic example of co-evolution, but the host–pollinator association is complex with many exceptions to host fidelity, exhibiting a reticulate, convoluted evolutionary history.

Sexual dimorphism is extreme, a function of the different ecological roles that the fig wasp sexes play in the interaction with their host figs and each other. Males are wingless and largely blind as they spend most of their short life as adults (they usually only live for hours) within the dark fig cavity, whereas females are winged as they need to fly off to find

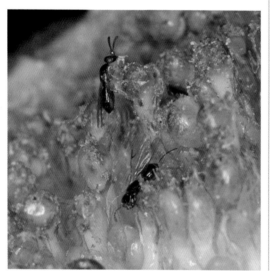

LEFT | Female pollinating fig wasps (*Ceratosolen capensis*) emerging from their natal galls within the central fig cavity of *Ficus sur.*

DISTRIBUTION
Worldwide, but restricted to warmer regions, most diverse in the tropics

GENERA
Agaon, Alfonsiella, Allotriozoon, Blastophaga, Ceratosolen, Courtella, Deilagaon, Dolichoris, Elisabethiella, Eupristina, Kradibia, Nigeriella, Paragaon, Pegoscapus, Platyscapa, Pleistodontes, Tetrapus, Waterstoniella, Wiebesia

HABITATS
Most species are associated with tropical forests or savanna ecosystems, but a few are adapted to more arid, temperate habitats

SIZE
On average $1/64$–$1/8$ in (0.5–2.5 mm) in length

ACTIVITY
Diurnal, crepuscular, or nocturnal as adult wasps, larvae resident within galls inside host figs

ABOVE | A pollinating fig wasp female (*Ceratosolen corneri*) about to enter the ostiole (an opening at the apex of the fig) of *Ficus botryocarpa* to effect pollination.

INSET | Underside of the head of a female pollinating fig wasp (*Agaon kiellandi*) showing the modified mandibular plate with rows upon rows of backward-pointing teeth that facilitate her negotiation of the ostiole.

REPRODUCTION
Females enter receptive figs through the ostiole to lay eggs into the ovules lining the fig cavity

DIET
Fig wasp larvae feed on endosperm tissue in the galled ovary. Adult wasps live for only a couple of days, or hours in the case of males, which do not feed as adults

receptive figs for pollination and oviposition. This remarkable feat is achieved by homing in on host tree-specific volatiles, a chemical signal released by the fig when it is receptive for pollination. Fig wasps provide a critical pollination ecosystem service, without which the host figs would not exist. Figs are keystone species in many ecosystems, providing a food resource for a wide variety of dependent animals.

APHELINID WASPS

Aphelinid wasps are a diverse worldwide family with 28 genera and 1,078 species in 5 subfamilies: Aphelininae, Calesinae, Coccophaginae, Eretmocerinae, and Eriaphytinae. However, recent comprehensive molecular and morphological analyses of the evolutionary relationships of Chalcidoidea have resulted in the elevation of Calesinae (containing the single genus *Cales*) to family level in the Chalcidoidea.

Species attack a wide variety of hosts in the order Hemiptera, most commonly mealybugs (Pseudococcidae), scale insects (Coccidae), whiteflies (Aleyrodidae), and aphids (Aphididae). Many genera are host-specific, with *Encarsia* and *Eretmocerus* only attacking whiteflies, and *Aphelinus* only aphids. Some attack Hemiptera, Lepidoptera, or Orthoptera eggs, while others are hyperparasitoids of other chalcid wasps. The majority of species are either external or internal parasitoids, but some species are known to exhibit divergent life history strategies across the sexes, where females develop as primary internal

LEFT | *Eretmocerus* parasitoids of whiteflies (Aleyrodidae).

ABOVE RIGHT | Close-up scanning electron microscope image of the head of *Aphelinus abdominalis*.

BELOW RIGHT | *Aphelinus abdominalis*, an effective biocontrol agent of pest aphid species attacking agricultural crops, such as the Potato Aphid (*Macrosiphum euphorbiae*, Aphididae).

DISTRIBUTION
Worldwide

GENERA
Allomymar, Aphelinus, Aphytis, Bardylis, Botryoideclava, Centrodora, Coccobius, Coccophagoides, Coccophagus, Dirphys, Encarsia, Eretmocerus, Eriaphytis, Eutrichosomella, Hirtaphelinus, Lounsburyia, Marietta, Marlattiella, Metanthemus, Oenrobia, Proaphelinoides, Prophyscus, Protaphelinus, Pteroptrix, Samariola, Timberlakiella, Verekia

HABITATS
Wide range, from forests to tundra

SIZE
$1/64–1/16$ in (0.5–1.5 mm) body length

ACTIVITY
Diurnal. Adults are rarely seen as they are so small, but can be easily swept or reared

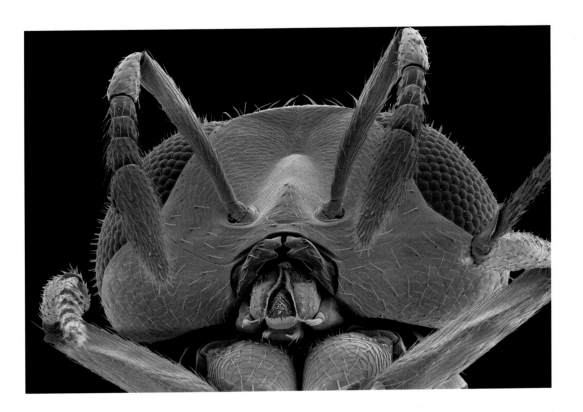

REPRODUCTION
Females lay eggs either into or onto
their hosts

DIET
Parasitoids of Hemiptera (Coccidae,
Aleyrodidae, Aphidae), or egg
parasitoids or hyperparasitoids of
Hemiptera, Lepidoptera, Orthoptera,
and Diptera

parasitoids and the males may develop on the same host as external parasitoids, or they may attack their own species or other chalcids as hyperparasitoids.

This family is critical in biocontrol programs, with large numbers of species having been used successfully to control pest species in the agricultural and forestry industries. *Aphelinus varipes* and *Aphidius colemani* successfully control aphids attacking vegetable crops; *Aphelinus mali* is a natural enemy of the Woolly Apple Aphid (*Eriosoma lanigerum*); and *Coccophagus scutellaris* controls a range of soft scale insect species (Coccidae) that are pests of citrus, guava, mango, and olive, to name a few. *Marietta javensis* is a hyperparasitoid of other aphelinid species controlling mealybug and scale pest of citrus and so has a detrimental effect on these programs.

Marietta species are very distinctive, being boldly marked in a black or brown and white zebra or piebald pattern.

AZOTID WASPS

This worldwide family contains a single genus, *Ablerus*, with 92 species, which until recently was included within the family Aphelinidae.

Ablerus species are primary parasitoids or hyperparasitoids of bug nymphs (Hemiptera), but some attack lepidopteran eggs or pupae of flies (Chamaemyiidae) that are predatory on aphids. *Ablerus macrochaeta* has been recorded attacking the Silverleaf Whitefly (*Bemisia tabaci*, Aleyrodidae), which is a globally destructive pest of vegetable and ornamental crops, as well as the Citrus Blackfly (*Aleurocanthus woglumi*), and the whitefly *Aleurocanthus inceratus*, which attacks sweet potato and tea in China. Together with a platygastrid and a couple of encyrtid species, *Ablerus perspeciosus* aids in the control of the Sugarcane Whitefly (*Aleurolobus barodensis*).

DISTRIBUTION
Worldwide

GENUS
Ablerus

HABITATS
A wide range, from forests to tundra

SIZE
$1/64$–$1/32$ in (0.5–1 mm) body length

ACTIVITY
Diurnal. Adults are rarely encountered

REPRODUCTION
Females lay eggs into the host, where her offspring may develop as primary parasitoids or as hyperparasitoids attacking parasitoid larva already developing within the host

DIET
Parasitoids or hyperparasitoids of insect eggs and larvae

BAEOMORPHID WASPS

This ancient family of rare wasps, previously known as Rotoitidae, comprises two living genera, each with a single species, one from Chile and one from New Zealand. Two extinct genera containing 14 species are only represented by Cretaceous fossils.

As part of a recent comprehensive molecular and morphological analysis of the evolutionary relationships of Chalcidoidea, Baeomorphidae was recognized as the senior synonym of Rotoitidae.

Wings of the Chilean species, *Chiloe micropteron*, are redundant, only present as tiny bristles, whereas females of the New Zealand species, *Rotoita basalis*, are fully winged.

The known distribution of the two living species suggests the family to be an archaic relict group of wasps. The amber fossil species are all known from the northern hemisphere and hence the family was considerably more diverse and globally widespread in the evolutionary past.

It is one of only a few parasitoid wasp families for which the hosts are completely unknown, but the association of adults with sphagnum moss has led to the conjecture that they may attack the eggs of the moss bug family Peloridiidae. Both species are associated with damp forest habitats. *Chiloe micropteran* has a dense coating of hairs that traps air around the body, enabling persistent breathing in the wet environments associated with mosses.

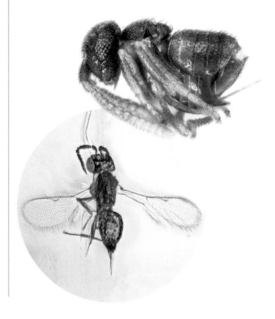

RIGHT | The apterous *Chiloe micropteron* (ABOVE) and the fossil species *Baeomorpha liorum* (RIGHT). Hosts are unknown for this family.

DISTRIBUTION
Chile, New Zealand

GENERA
Chiloe, Rotoita

HABITATS
Temperate forests and grassland

SIZE
$^1/_{32}$–$^1/_{16}$ in (1–2 mm) body length

ACTIVITY
Adults are rarely encountered

REPRODUCTION
Unknown

DIET
Unknown

CHALCID WASPS

BELOW | *Brachymeria ovata*, a polyphagous species attacking many species of butterflies and moths (Lepidoptera).

BOTTOM | *Conura*, a species-rich genus most diverse in the Neotropical region.

Chalcidid wasps are a worldwide family containing 83 genera and 1,548 described species placed in 8 subfamilies: Brachymeriinae, Chalcidinae, Cratocentrinae, Dirhininae, Epitraninae, Haltichellinae Phasgonophorinae, and Smicromorphinae.

These wasps are parasitoids of the larvae or pupae of butterflies and moths (Lepidoptera) or flies (Diptera), less commonly other Hymenoptera, lacewings, antlions (Neuroptera), or beetles (Coleoptera).

Chalcididae are usually black or brown, with many having yellow or red patches on the legs. They are easily recognizable by their enlarged hind femur, which often has teeth along the ventral margin.

DISTRIBUTION
Worldwide but with highest species richness in the tropics

GENERA
83 genera in 8 subfamilies: Brachymeriinae, Chalcidinae, Cratocentrinae, Dirhininae, Epitraninae, Haltichellinae, Phasgonophorinae, and Smicromorphinae.

The four genera *Antrocephalus*, *Brachymeria*, *Conura*, and *Hockeria* include more than half of the described species

HABITATS
Wide range, from forests to tundra

SIZE
$1/16$–$9/16$ in (2–15 mm) body length

ACTIVITY
Diurnal

Phasgonophora sulcata, a
parasitoid of pest jewel beetle larvae
developing in tree trunks, such as the
Two-lined Chestnut Borer (*Agrilus
bilineatus*) and the Emerald Ash Borer
(*Agrilus planipennis*), both in the
family Buprestidae.

REPRODUCTION
Females lay eggs directly into the host,
usually larvae

DIET
As parasitoids the larvae feed on the
internal tissue of their host insects

Females of *Lasiochalcidia igiliensis* use their large hind
legs to hold open the massive, predatory mandibles of
antlion larva so that they can lay an egg into the host's
exposed throat membrane.

Cratocentrinae have long ovipositors evolved to reach
their host wood-boring beetle larvae feeding deep in trunks
or tree branches.

Dirhininae species have characteristic horn-like
projections on their heads—they attack flies.

Epitranus species have a characteristically elongate petiole
and are parasitoids of mostly moths (Lepidoptera), but some
species are myrmecophiles or termitophiles living in ant or
termite nests.

CHRYSOLAMPID WASPS

Chrysolampid wasps are a recently erected worldwide family containing 9 genera and about 75 species in 2 subfamilies: Chrysolampinae and Philomidinae.

Chrysolampinae species attack beetles (Curculionidae, Nitidulidae, and possibly Cerambycidae). The single host association for Philomidinae suggests that they are parasitoids of the larvae of ground-nesting bees (Halictidae), emerging from the pupa. The first-instar planidial larva of *Aperilampus varians* attacks the pupa of the bee *Halictus africanus*. *Philomides* species have been recorded ovipositing into flower buds of *Solanum* (Solanaceae), suggesting that the mobile first-instar larvae then attach to bees visiting the flowers for nectar and pollen, thereby hitching a ride to the underground bee nest. *Chrysolampus thenae* is an external parasitoid of the beetle *Meligethes pedicularius* (Nitidulidae) living in the flower heads of Betony (*Betonica officinalis*).

LEFT | *Aperilampus*, parasitoids of ground-nesting bees.

RIGHT | A female *Cynipencyrtus flavus* reared from a gall of the Oriental Chestnut Gall Wasp in Japan.

FAR RIGHT | A leaf gall of the Oriental Chestnut Gall Wasp (*Dryocosmus kuriphilus*, Cynipidae), that is a host of *Cynipencyrtus flavus*.

DISTRIBUTION
Worldwide

GENERA
Aperilampus, Australotoxeuma, Brachyelatus, Chrysolampus, Chrysomalla, Elatomorpha, Parelatus, Philomides, Vidlinus

HABITATS
A wide range, from forests to tundra

SIZE
$1/16$–$3/16$ in (1.5–5 mm) body length

ACTIVITY
Diurnal. Adults are rarely encountered

REPRODUCTION
Females probably lay eggs near their host given that they have mobile first-instar planidial larvae

DIET
Wasp larvae feed on body tissue of the host beetle or bee larvae

CYNIPENCYRTID WASPS

This rare family has a single genus and species, *Cynipencyrtus flavus*, which is only known from China and Japan.

Cynipencyrtidae is a fairly recently erected family to accommodate the genus *Cynipencyrtus*, which was originally placed in the Encyrtidae and subsequently transferred to the Tanaostigmatidae. Comprehensive analyses of the evolutionary relationships in Chalcidoidea in 2013 supported the recognition of this evolutionary lineage as a distinct family.

In Japan it is a parasitoid of the Oriental Chestnut Gall Wasp (*Dryocosmus kuriphilus*, Cynipidae), which is a global pest of chestnut (*Castanea*). The species also attacks several species of *Andricus* (Cynipidae) that form galls on *Quercus serrata* (Fagaceae).

DISTRIBUTION
China and Japan

GENUS
Cynipencyrtus

HABITATS
Temperate forests

SIZE
¹/₁₆ in (1.5 mm) body length

ACTIVITY
Diurnal. Adults are rarely encountered

REPRODUCTION
Females lay eggs into the host gall containing the developing gall-former on chestnut or oak

DIET
The parasitoid larvae feed on the body contents of their host gall-forming wasp

ENCYRTID WASPS

Encyrtid wasps are a diverse and species-rich worldwide family with about 460 genera and around 3,750 described species in 2 subfamilies, Encyrtinae and Tetracneminae. This is one of the most morphologically diverse of all hymenopteran families, with many bizarre and colorful species, and a huge undescribed diversity.

Most species and all of Tetracneminae are parasitoids of Homopteran bugs (Coccidae, Diaspididae, Pseudococcidae), but a wide variety of other insects and arachnids is also targeted by species of Encyrtinae: eggs or larvae of beetles (Coleoptera), flies (Diptera), butterflies and moths (Lepidoptera), lacewings (Neuroptera), grasshoppers (Orthoptera), bugs (Heteroptera), and spiders (Araneae), as well as nymphs of ticks (Ixodida). Some species are parasitoids or hyperparasitoids of other wasps (Hymenoptera).

BELOW | *Ericydnus strigosus,* a parasitoid of the mealybug *Heterococcus pulverarius* (Pseudococcidae).

DISTRIBUTION
Worldwide

GENERA
460 genera in 2 subfamilies, Encyrtinae and Tetracneminae

HABITATS
A wide range, from forests to tundra

SIZE
$1/64$–$3/16$ in (0.5–4 mm) body length

ACTIVITY
Diurnal. Adults are rarely encountered but can be swept or reared in numbers

REPRODUCTION
Females lay eggs directly into their host insect or arachnid

DIET
The larvae feed on the body contents of their host insect or spider they are developing on

ABOVE | *Encyrtus aurantii*, an
effective biocontrol agent of soft
scale species (Coccidae) attacking
subtropical fruit trees.

BELOW | *Bothriothorax*, parasitoids
of hoverflies (Syrphidae), or
hyperparasitoids of caterpillars
(Lepidoptera) via bristle flies
(Tachinidae).

Encyrtids are all internal parasitoids of their
hosts. A few are polyembryonic: *Copidosoma* females
lay a single egg, which divides and multiplies into
hundreds or in some cases thousands of cloned eggs
within a single caterpillar, all resulting in genetically
identical adults. Some of the larvae act as "soldiers"
within the host body and do not develop into
adulthood.

Many species are important biocontrol agents
of pest insects in the agricultural and forestry
industries. *Aenasius bambawalei* is a primary
parasitoid that controls populations of the
Cotton Mealybug (*Phenacoccus solenopsis*,
Pseudococcidae). *Metaphycus macadamiae*
successfully controls the Macadamia
Felted Coccid (*Acanthococcus
ironsidei*, Eriococcidae), in Hawaii.

127

Anagyrus pseudococci and *Leptomastix dactylopii* are efficient biocontrol agents attacking *Planococcus citri*, a mealybug pest in citrus orchards worldwide. *Comperiella bifasciata* controls Red Scale (*Aonidiella aurantii*), another economically important pest of citrus.

Encyrtids have been some of the most economically important biological control agents, with *Apoanagyrus lopezi* successfully saving Cassava crops in Africa and Southeast Asia from the Cassava Mealybug (*Phenacoccus manihoti*). Indirectly, *Tachardiaephagus somervillei* could save Christmas Island crabs by controlling the invasive scale insect *Tachardina aurantiaca*, the honeydew of which is a food source for another invasive species, the Yellow Crazy Ant (*Anoplolepis gracilipes*), which attacks the crabs.

TOP | *Metaphycus*, a species-rich genus with over 250 species that mostly attack soft scales (Coccidae).

ABOVE | *Cowperia areolata*, a parasitoid of ladybird beetles (Coccinellidae).

ERIAPORID WASPS

This fairly recently erected family was previously classified as a subfamily in the Aphelinidae. The Eriaporidae are naturally distributed only in the Old World (i.e., excluding the Americas, although *Myiocnema comperei* was introduced to California for biological control), with 5 genera and 22 species in 2 subfamilies: Eriaporinae and Euryischiinae. As we write, the family Eriaporidae is in the process of being synonymized with a newly established family, Pirenidae, with Eriaporinae and Euryischiinae retained as subfamilies. The scientific paper implementing these formal classification changes will have been published by the time this book goes to press.

Eriaporid wasps are parasitoids or hyperparasitoids of insect eggs and larvae and are of biocontrol significance, attacking pest species of insects or having a negative impact by attacking the primary parasitoid of the pests. Species of *Eunotiscus* are parasitoids of *Pseudococcus* (Pseudococcidae). *Euryischia indica* controls the Seychelles Scale Insect (*Icerya seychellarum*). *Myiocnema comperei* and *Promuscidea unfasciativentris* attack the primary parasitoid *Aenasius bambawalei* (Encyrtidae) that controls populations of the Cotton Mealybug (*Phenacoccus solenopsis*, Pseudococcidae), developing as a pest on cotton.

LEFT | *Eunotiscus*, parasitoids of mealybugs (Pseudococcidae).

DISTRIBUTION
Africa, Asia, Australia

GENERA
Eunotiscus, Euryischia, Euryischomyia, Myiocnema, Promuscidea

HABITATS
A wide range, from forests to tundra

SIZE
$^1/_{64}$–$^1/_{16}$ in (0.5–2 mm) body length

ACTIVITY
Diurnal. Adults are rarely encountered

REPRODUCTION
Females lay eggs into or onto their hosts

DIET
Parasitoid larvae feed on body contents of their host insect eggs and larvae

EUCHARITID WASPS

These wasps are a worldwide family with 57 genera and 427 species placed in 4 subfamilies: Akapalinae, Eucharitinae, Gollumiellinae, and Oraseminae.

All species of eucharitid wasps are parasitoids of ants. Huge numbers of eggs (up to 15,000) are laid on vegetation. Eucharitids have a free-living, mobile first-instar larva, the "planidium" (meaning "diminutive wanderer"), which attaches itself to ants foraging on the plants and in this process is carried back to the ant's nest. It waits until the ant larva pupates and then commences feeding as an ectoparasitoid. The planidium will sometimes feed on an intermediate carrier such as a thrips (Thysanoptera) if it is initially unsuccessful in attaching to an ant worker when it hatches.

Psilocharis females lay their eggs among trichomes under the bracts present at the base of flowers, instead of inserting their eggs into slits or cavities in the plant tissue as is normal for the rest of the family.

Orasema species are potential biocontrol agents against the invasive Red Imported Fire Ant (*Solenopsis invicata*) and the invasive Little Fire Ant (*Wasmannia auropunctata*, Formicidae).

LEFT | *Stilbula cyniformis*, a parasitoid of ants in the genus *Camponotus* (Formicidae).

DISTRIBUTION
Worldwide, but absent from colder regions and most diverse in tropical and subtropical regions

GENERA
57 genera. Some of the more common include *Akapala, Ancylotropus, Babcokiella, Cherianella, Eucharis, Eucharissa, Gollumiella, Hydrorhoa, Kapala, Mateucharis, Neostilbula, Orasema, Pseudochalcura, Psilocharis, Stilbula, Timioderus*

HABITATS
A wide range, from forests to grassland

SIZE
$^1/_{32}$–$^3/_{16}$ in (1–5 mm) body length

ACTIVITY
Diurnal. Adults are infrequently encountered

ABOVE | *Eucharis*, parasitoids of ants in the subfamilies Formicinae and Ponerinae (Formicidae).

RIGHT | *Stilbula* male.

REPRODUCTION
Females lay their eggs into plants. The mobile planidium larva attaches to foraging ant workers to hitch a ride into the nest

DIET
Parasitoids of ant larvae, feeding on the host's body tissue

EULOPHID WASPS

BELOW | *Tetrastichus planipennisi*, an Asian species introduced to North America as a biocontrol agent of the Emerald Ash Borer (*Agrilus planipennis*, Buprestidae), an introduced, invasive species decimating the American ash forests.

This is a worldwide family with 334 genera and about 5,000 species placed in 5 subfamilies: Entedoninae, Entiinae, Eulophinae, Opheliminae, and Tetrastichinae. The genus *Elasmus* was previously placed in its own family, Elasmidae.

These wasps are external or internal parasitoids of a wide range of insect orders, but mainly Lepidoptera, Coleoptera, Hymenoptera, and Diptera. A few species are phytophagous, and some are predators of gall mites or even nematodes.

The overall host range and diversity of life history strategies of eulophid wasps is immense, and this is a very species-rich family, with most tropical species awaiting description.

Many species are of economic importance, acting as biological control agents of pest species. *Neochrysocharis formosa* and *Diglyphus begini* are important natural enemies of various *Liriomyza* leaf-miner fly species (Agromyzidae) that attack vegetables. *Aprostocetus* species attack the Blueberry Gall Midge complex (Cecidomyiidae) and help control the Sugarcane Planthopper (*Perkinsiella saccharicida*,

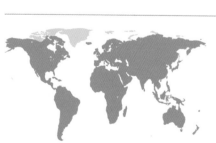

DISTRIBUTION
Worldwide

GENERA
334 genera. Some of the more common include *Aprostocetus, Astichus, Baryscapus, Ceranisus, Cirrospilus Closterocerus, Diglyphus, Elasmus, Entedon, Euderus, Euplectrus, Neochrysocharis, Neotrichoporoides,*

Oomyzus, Ophelimus, Pediobius, Quadrastichodella, Quadrastichus, Sympiesis, Tamarixia, Tetrastichus, Trichospilus

HABITATS
A wide range, from forests to tundra

SIZE
$1/64$–$1/4$ in (0.4–6 mm) body length

ACTIVITY
Diurnal. Adults are commonly encountered

REPRODUCTION
Females lay their eggs into or onto their host

DIET
Larvae feed on body contents of their host insects

Delphacidae). Species of *Melittobia* are parasitoids of solitary bees and wasps and also their nest co-inhabitants. *Ceranisus* species attack the nymphs of *Magalurothrips*, a pest of pulses. *Entedon* species attack bark beetles. *Pediobius* species control skipper butterfly caterpillars feeding on banana crops. *Entedon magnificus* is a natural enemy of *Gonipterus* weevils (Curculionidae), which are pests in eucalyptus plantations. *Closterocerus oryzamyntor* attacks larvae of the pest Rice Hispa Leaf Beetle (*Dicladispa armigera*, Chrysomelidae). *Tamarixia radiata* is an important biocontrol agent of the Asian Citrus Psyllid (*Diaphorina citri*, Psyllidae), which, as the vector for the economically impactful citrus greening disease, is a serious citrus pest.

EUPELMID WASPS

BELOW | *Balcha* female ovipositing into a wood-boring jewel beetle larva (Buprestidae) feeding inside a tree branch.

With a worldwide distribution, this family previously had 51 genera and about 931 species placed in 3 subfamilies: Calosotinae, Eupelminae, and Neanastatinae. Recent comprehensive molecular and morphological analyses of the evolutionary relationships of Chalcidoidea have resulted in the elevation of Neanastatinae (containing two genera:

©2019 melvyn yeo

DISTRIBUTION
Worldwide

GENERA
49 genera in two subfamilies: Calosotinae, Eupelminae. A few of the more common genera include: *Anastatus, Arachnophaga, Calosota, Euplemus, Eusandalum, Mesocomys, Tineobius*

HABITATS
A wide range, from forests to tundra

SIZE
$1/32–3/16$ in (1–5 mm) body length

ACTIVITY
Diurnal. Adults are frequently encountered but at low density

REPRODUCTION
Females lay their eggs into or onto their host

DIET
Larvae feed on egg contents or body contents of their host insects

Lambdobregma and *Neanastatus*) to family level in the Chalcidoidea, and the genus *Metapelma* was transferred to the newly erected family Metapelmatidae (a summary list of these new families is provided on page 154).

Eupelmids are parasitoids or pseudohyperparasitoids of insects whose larvae are usually developing in a concealed situation, or parasitoids of insect or spider eggs. Species of Calosotinae usually attack wood-boring beetles, but the common species *Calosota metallica* is a parasitoid of a range of different insects developing in grass stems. Two very common species, *Eupelmus urozonus* and *Eupelmus vesicularis*, can either be primary external parasitoids or pseudohyperparasitoids, attacking a diverse range of other primary parasitoids.

The Eupelminae have evolved the ability to jump vigorously via the contraction of large muscles that contort and contract the thorax, suddenly releasing the stored energy and resulting in the midlegs kicking out strongly.

TOP | *Anastatus semiflavidus*, an egg parasitoid of the Range Caterpillar (*Hemileuca oliviae*, Saturniidae), leaf-footed bugs (Coreidae), and stink bugs (Pentatomidae).

ABOVE | *Anastatus* female selecting suitable host eggs for oviposition.

EURYTOMID WASPS

This common, worldwide family comprises 97 genera and 1,453 described species placed in 4 subfamilies: Buresiinae, Eurytominae, Heimbrinae, and Rileyinae.

Eurytomid wasps have three basic life histories: phytophagous (plant-feeding) as seed feeders, stem feeders, or gall-formers; parasitoids of a wide range of insects; and inquilines, feeding on both the host insect and surrounding plant tissue.

The seed eaters include two genera: *Systole*, which feed on Umbelliferae seeds, and *Bruchophagus*, which are often reared from seeds of plants in the Leguminosae. *Bruchophagus gibbus* can be a pest species affecting crops of White Clover (*Trifolium repens*). Stem feeders include species of the common genus *Tetramesa*, which may become pests of cereal crops. Many species of *Sycophila* are obligate associates of *Ficus* species, attacking Epichrysomallidae developing in galled florets within the fig cavity.

DISTRIBUTION
Worldwide

GENERA
97 genera. Some of the more common or widespread include *Bruchophagus*, *Eurytoma*, *Philolema*, *Prodecatoma*, *Rileya*, *Sycophila*, *Systole*, *Tetramesa*

HABITATS
A wide range, from forests to tundra

SIZE
$^1/_{32}$–$^3/_{16}$ in (1–5 mm) body length

ACTIVITY
Diurnal. Adults are commonly encountered

REPRODUCTION
Females lay their eggs into or onto their host

DIET
Larvae feed on egg contents, or the body contents of their host insects, or on plant tissue

ABOVE LEFT | *Eurytoma* male mounting a female before the female oviposits into a *Bulbine* seed capsule (Xanthorrhoeaceae). Her larvae will feed on the seeds, destroying them in the process.

ABOVE | *Eurytoma* ovipositing into a microgastrine (Braconidae) cocoon spun on the outside of the host caterpillar, an example of hyperparasitism.

Species in the subfamily Rileyinae are often predators of eggs of other insects. *Eurytoma solenozophaeriae* is associated with the Blueberry Stem Gall Wasp (*Hemadas nubilipennis*), galling Lowbush Blueberry (*Vaccinium angustifolium*).

EUTRICHOSOMATID WASPS

Eutrichosomatid wasps are a rarely encountered family, recently erected for three genera with five species previously placed in the Pteromalidae: *Collessina* (Australia), *Eutrichosoma* (North America and Brazil), and *Peckianus* (North America and Brazil). There are three species of *Eutrichosoma*: *E. mirabile*, *E. flabellatum*, and *E. burksi*, and one species of each of the other two genera.

Based on the few host records, these wasps attack seed-feeding weevils. *Eutrichosoma mirabile* is a parasitoid of weevils in the genera *Auleutes* and *Smicronyx* (Curculionidae) that feed on *Parthenium* and *Helianthus* (Asteraceae). *Peckianus laevis* is a parasitoid of weevils in the genus *Apion* (Brentidae). The host of the Australian *Collessina pachyneura* is unknown.

These wasps have active first-instar larvae called "planidia" (*see also* page 130). Eggs are usually laid near to the host rather than directly into or onto the host, which means the larvae that hatch need to be mobile to find the host. They feed externally and need to detach and reattach as the host molts between instars. The first instar undergoes metamorphosis to develop into the typical non-mobile parasitoid second-instar larva, a transition that only occurs when the host larva has pupated.

LEFT | *Eutrichosoma mirabile*, a parasitoid of seed-feeding weevils (Curculionidae).

RIGHT | *Leucospis gigas*, a parasitoid of mortar bees (Megachilidae) and paper wasp larvae (Vespidae).

DISTRIBUTION
Australia, Brazil, and North America

GENERA
Collessina, Eutrichosoma, Peckianus

HABITATS
Forests, grassland

SIZE
$^1/_{32}$–$^1/_{16}$ in (1–2 mm) body length

ACTIVITY
Diurnal. Adults are rarely encountered

REPRODUCTION
Females lay eggs in the vicinity of the host beetle eggs, usually within seed pods

DIET
The parasitoid wasp larvae feed externally on weevil host larvae and pupae

LEUCOSPID WASPS

This worldwide family of usually rather uncommon species has 4 genera with 134 described species.

They are external parasitoids of solitary bees (Apidae) and wasps (Eumeninae and Sphecidae). *Leucospis gigas* is a large species, whose eggs are ¹⁄₈ in (3 mm) long, exceptionally large for a chalcid.

Oviposition takes place by penetration of the nest cell wall of the host, which could be mud or wood. The ovipositor has a unique structure in that it is housed in a longitudinal notch on top of the gaster.

These wasps mimic bees and stinging wasps in color patterns, often being black with yellow, white, or red markings. This warning coloration helps to reduce predation by vertebrates. They have a strongly swollen, toothed hind femur.

The South American *Leucospis pinna* has gregarious larvae attacking the Orchid Bee (*Eulaema meriana*, Apidae).

DISTRIBUTION
Worldwide, predominantly in tropical and subtropical regions, but absent from New Zealand

GENERA
Leucospis, Micrapion, Neleucospis, Polistomorpha

HABITATS
A wide range, from forests to tundra

SIZE
¹⁄₁₆–⁵⁄₈ in (2–16 mm) body length

ACTIVITY
Diurnal. Adults are rarely encountered

REPRODUCTION
Females lay their eggs onto or nearby the host wasp or bee larva

DIET
Larvae feed externally on body contents of their host insects. Adults take nectar at flowers

MEGASTIGMID WASPS

The megastigmid wasps are a worldwide family previously with 14 genera and more than 200 described species. Recent comprehensive molecular and morphological analyses of the evolutionary relationships of Chalcidoidea resulted in the transfer of two subfamilies from Pteromalidae to Megastigmidae: Chromeurytominae and Keiraninae, adding another four genera to the family.

Most megastigmids are phytophagous, feeding on seeds of Pinaceae and Rosaceae, but some species in the genera *Bootanomyia* and *Megastigmus* are external parasitoids of gall wasps (Cynipidae), and those in *Mangostigmus*, *Megastigmus*, and *Neomegastigmus* attack gall-forming flies (Cecidomyiidae).

LEFT | *Megastigmus* female, with typical curved ovipositor sheaths.

DISTRIBUTION
Worldwide

GENERA
Asaphoideus, Bootanelleus, Bootania, Bootanomyia, Bortesia, Chromeurytoma, Ianistigmus, Keirana, Macrodasyceras, Malostigmus, Mangostigmus, Megastigmus, Neomegastigmus, Paramegastigmus, Patiyana, Striastigmus, Vitreostigmus, Westralianus

HABITATS
A wide range, from forests to tundra

SIZE
$1/32$–$1/4$ in (1–7 mm) body length

ACTIVITY
Diurnal. Adults are commonly encountered around their host plant

REPRODUCTION
Females lay their eggs into plant tissue

DIET
Phytophagous, feeding on tissue of seeds or plants, or parasitoids of gall-forming insects

Species of the Australian genus *Bortesia* are gall-formers in buds of *Hakea* species. *Megastigmus* species that feed on conifer seeds can be important economic pests, such as the Larch Seed Chalcid (*Megastigmus pictus*). The Rose-Hip Chalcid (*Megastigmus aculeatus*) is a seed predator of roses, damaging seeds of both indigenous and naturalized species, and hence impacts the rose hip industry, but may also be beneficial as a biocontrol agent of the invasive Sweet Brier (*Rosa rubiginosa*).

The Brazilian Peppertree Seed Chalcid (*Megastigmus transvaalensis*), an African gall-former species developing in seed capsules of *Searsia* (=*Rhus*) species (Anacardiaceae), has host-shifted to the Peruvian Peppertree (*Schinus molle*) and the Brazilian Peppertree (*Schinus terebinthifolia*) (Anacardiaceae), which are invasive in many parts of the world. This wasp has beneficial impact where it has established in invasive populations of pepper trees in the USA, but has also now reached Brazil, where it has negative impact on the indigenous populations of these trees in South America.

FAIRYFLY WASPS

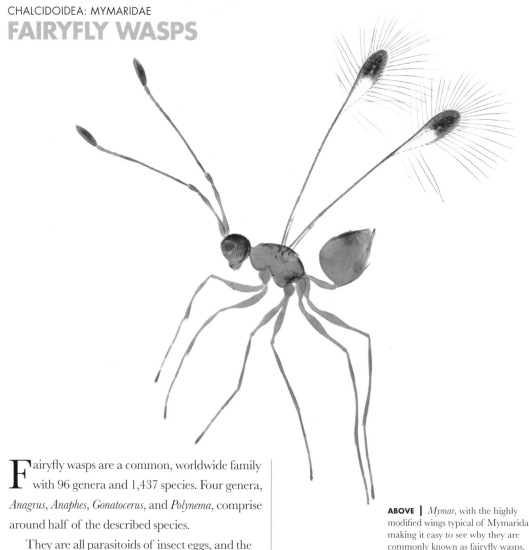

Fairyfly wasps are a common, worldwide family with 96 genera and 1,437 species. Four genera, *Anagrus*, *Anaphes*, *Gonatocerus*, and *Polynema*, comprise around half of the described species.

They are all parasitoids of insect eggs, and the majority of species are tiny, a function of the very limited available food resource within the host insect egg that the larvae feed on.

ABOVE | *Mymar*, with the highly modified wings typical of Mymaridae, making it easy to see why they are commonly known as fairyfly wasps.

ABOVE RIGHT | Female mymarids have obvious clubbed antennae, whereas males have uniformly thread-like antennae.

DISTRIBUTION
Worldwide

GENERA
96 genera. Some of the more common include *Alaptus*, *Anagrus*, *Anaphes*, *Camptoptera*, *Erythmelus*, *Gonatocerus*, *Litus*, *Mymar*, *Ooctonus*, *Palaeoneura*, *Polynema*, *Stethynium*

HABITATS
A wide range, from forests to tundra

SIZE
$^1/_{64}$–$^1/_8$ in (0.15–2.5 mm) body length

ACTIVITY
Diurnal. Adults are common, but rarely noticed because of their size

REPRODUCTION
Females lay their egg directly into the host egg

DIET
Larvae feed on the contents of their host egg

The smallest recorded adult insect is a fairyfly, *Dicopomorpha echmepterygis*, where males are wingless, and only 0.127 mm long. Their bodies are reduced and simplified, as they never leave the host egg, mating with their sisters before they make their way into the great big world. The smallest recorded winged insects are females of the fairyfly *Kikiki huna* (1/64 in/0.15–0.19 mm). Another tiny fairyfly species has the apt name *Tinkerbella nana*.

Mymarids have long wing fringes, a characteristic often typical of the smallest insects, which almost "swim" through the air. *Polynema sagittaria* has a bizarre morphological adaptation to house the ovipositor—a bow-like structure extending from the abdomen forward beneath the body and projecting between the front legs. This species also has wings covered with strong bristles and a long fringe of setae, the latter characteristic for the family. *Mymarilla wollastoni*, a species endemic to the island of Saint Helena in the southern Atlantic, has extremely densely hairy and domed forewings, probably an adaptation to the species' existence on an island with inhospitable environmental conditions.

ORMYRID WASPS

BELOW | *Ormyrus nitidulus*, a parasitoid of oak gall wasps commonly of the genus *Andricus* (Cynipidae), and more rarely gall midges (Cecidomyiidae).

A worldwide family containing 3 genera and about 145 described species, ormyrid wasps are parasitoids of gall-forming insects, which are often other wasp species. The female lays her eggs through the gall wall into the gall-forming insect larva. The parasitoid wasp larva then eats the host gall insect larva.

Ormyrulus gibbus is a parasitoid of gall midges (Cecidomyiidae). Species of *Ormyrus* are primary parasitoids of gall wasps (Cynipidae), or Tephritidae (Diptera), or other gall-forming chalcid wasps in the families Agaonidae, Eulophidae, Eurytomidae, and Pteromalidae, or weevils (Curculionidae). *Ormyrus rosae* is a parasitoid of cynipid galls on roses and pteromalid galls on blueberries (*Vaccinium*) (Ericaceae). *Ormyrus vacciniicola* attacks the stem galler *Hemadas nubilipennis* developing on Lowbush Blueberry (*Vaccinium angustifolium*). A few species are parasitoids of fig wasps developing in figs (*Ficus*). *Ormyrus labotus* is a parasitoid of at least 65 gall wasps galling oaks, but this species is actually a complex of 16–18 cryptic species, so the host range will be narrower per species.

DISTRIBUTION
Worldwide

GENERA
Eubeckerella, Ormyrulus, Ormyrus

HABITATS
A wide range, from forests to tundra

SIZE
$^1/_{32}$–$^3/_{16}$ in (1–5 mm) body length

ACTIVITY
Diurnal. Some species can be collected readily from trees such as oaks (*Quercus*)

REPRODUCTION
Females lay their eggs into the gall cavity containing the primary gall-former

DIET
Larvae feed on their host gall-forming insect developing inside galls

ABOVE | *Ormyrus* male, associated with figs of *Ficus burkei* in Africa.

RIGHT | The vivid coloration of an *Ormyrus* metasoma.

0.50 mm

PERILAMPID WASPS

This worldwide family of wasps, with 6 genera and about 200 species, are often brilliantly colored metallic blue or green, sometimes black.

Perilampids are primary or hyperparasitoids of Diptera, Coleoptera, and Hymenoptera. Species of *Euperilampus* attack tachinid flies and Darwin wasps parasitizing Lepidoptera and emerge from the pupae of their secondary hosts. They are either hyperparasitoids of Lepidoptera through parasitoid flies in the family Tachinidae, or of parasitoid wasps in the superfamily Ichneumonoidea, or they are primary parasitoids of wood-boring beetle larvae (Anobiidae and Platypodidae), and less commonly Hymenoptera, Orthoptera, or Neuroptera.

As with Chrysolampidae and Eutrichosomatidae, these species have an active first-instar planidial larva that feeds externally and only attacks the host when it pupates. The hyperparasitoid species find a host caterpillar that is parasitized by a fly or wasp, entering the body of the primary parasitoid that is feeding inside the caterpillar, but then waiting without feeding for the primary parasitoid to finish development and pupate, at which time the perilampid larva exits to feed externally on the pupae.

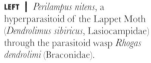

LEFT | *Perilampus nitens*, a hyperparasitoid of the Lappet Moth (*Dendrolimus sibiricus*, Lasiocampidae) through the parasitoid wasp *Rhogas dendrolimi* (Braconidae).

DISTRIBUTION
Worldwide

GENERA
Burskilampus, Euperilampus, Krombeinius, Monacon, Perilampus, Steffanolampus

HABITATS
A wide range, from forests to tundra

SIZE
1/16–1/4 in (1.5–6 mm) body length

ACTIVITY
Diurnal. Adults are found at low density

REPRODUCTION
Females probably lay eggs near their host, given that they have mobile first-instar planidial larvae

DIET
Wasp larvae feed either on body tissue of the host beetle larvae, or on fly or wasp larvae that are parasitizing caterpillars

Adults feed at flowers, some feed on aphid honeydew, and *Perilampus aeneus* has been observed feeding on plant exudate after puncturing the epidermis of a leaf. *Perilampus hyalinus* can either attack primary parasitoids developing in caterpillars of the Fall Webworm Moth (*Hyphantria cunea*, Erebidae), or develop as a primary parasitoid of sawflies (Diprionidae). *Perilampus similis* attacks braconid wasps, including *Agathis metzneriae*, a primary parasitoid of caterpillars of the moth *Metzneria lappella* (Gelechiidae), which feeds on burdock seeds.

TOP | *Perilampus prothoracicus*, a hyperparasitoid of moths in the families Oecophoridae, Pyralidae, and Tortricidae, probably via Darwin or braconid wasps.

INSET | *Perilampus hyalinus*, a primary or hyperparasitoid of a wide range of insects.

PTEROMALID WASPS

LEFT | *Pteromalus* female, a typical example of the ground plan morphology of Pteromalidae, whereas those associated with specialized niches such as figs have evolved some extreme morphological adaptations as shown on page 151.

RIGHT | *Mesopolobus tibialis*, associated with plant galls, mostly those of gall wasps on oaks (Cynipidae).

Previously, this was a huge and diverse, worldwide family with about 640 genera and about 3,550 described species placed in 33 subfamilies, but it has now been dramatically reduced in circumscription. Historically this has been in part a "dustbin" family receiving genera and subfamilies that taxonomists did not know where to place in the classification, as well as "core" pteromalids.

Recent comprehensive molecular and morphological analyses of the evolutionary relationships of Chalcidoidea have resulted in the elevation of 23 of the former pteromalid subfamilies or tribes to family level in the Chalcidoidea (a summary of these new families is provided on pages 152–54). Other subfamilies have been removed from Pteromalidae and transferred to other families: Chromeurytominae and Keiraninae to Megastigmidae; Elatoidinae to Neodiparidae; Nefoeninae to Pelecinellidae; and Erotolepsiinae to Spalangiidae. This leaves eight currently recognized subfamilies in the Pteromalidae: Colotrechninae, Erixestinae, Miscogastrinae, Ormocerinae, Pachyneurinae, Pteromalinae, Sycophaginae, and Trigonoderinae.

The non-pollinating fig wasps in the subfamily Sycophaginae, previously included in the Agaonidae, have been moved back to the Pteromalidae. The other three subfamilies of non-pollinating fig wasps (Otitesellinae, Sycoecinae,

DISTRIBUTION
Worldwide

GENERA
425 genera in eight subfamilies

HABITATS
A wide range, from forests to tundra

SIZE
$^1/_{64}$–$^1/_4$ in (0.5–6 mm) body length

ACTIVITY
Diurnal. Adults are usually readily collected by sweeping

REPRODUCTION
Females lay their eggs directly into or onto the host insect or egg

DIET
The larvae feed on the contents of their host insect

and Sycoryctinae) have been reduced in rank and are now assigned to the Pteromalinae in the single tribe Otitesellini.

Life history strategies of pteromalids are diverse. Species may be primary parasitoids, hyperparasitoids, inquilines, or kleptoparasitoids of a wide variety of insects and arachnids, or phytophagous gall-formers. Colotrechinae and Ormocerinae are usually associated with galls, and some are parasitoids of beetles (Coleoptera). Erixestinae are egg parasitoids of leaf-beetles (Chrysomelidae). Miscogastrinae are parasitoids of leaf-mining or stem-burrowing flies (Diptera), and rarely fleas (Siphonaptera). Pachyneurinae are parasitoids of giant scales (Hemiptera), some of which are important agricultural pests.

Pteromalinae contain a wide range of host associations from parasitoids or hyperparasitoids of many insect orders, predators within spider egg sacs to gall-formers. Sycophaginae are specialized gall-formers within figs. Trigonoderinae are parasitoids of wood-boring beetles.

The only known parasitoid of flea larvae is the pteromalid *Bairamlia fuscipes*. One pteromalid, *Nasonia vitripennis*, in nature a parasitoid of fly puparia in bird nests, has become a model organism for study in the laboratory, particularly in the fields of sex allocation and developmental genetics.

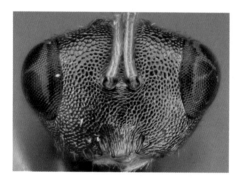

LEFT | Head of a male *Pteromalus puparum*, a ubiquitous, cosmopolitan, gregarious internal parasitoid of butterfly pupae. Females oviposit directly into the soft, freshly formed pupal cuticle.

RIGHT | *Philotrypesis*, a parasitoid non-pollinating fig wasp that lays eggs from the outside of the fig, using her long ovipositor to reach the host fig wasps developing in galled florets inside the fig.

BELOW | A pteromalid female attacking microgastrine (Braconidae) larvae/pupae in their cocoons, the larvae having erupted from their host caterpillar to spin and pupate.

NEW FAMILIES
OF CHALCIDOIDEA

While this book was in the process of final production the results of a comprehensive molecular and morphological analysis of the evolutionary relationships of Chalcidoidea were published, which has resulted in the recognition and establishment of 26 new wasp families. The "dustbin" family Pteromalidae (see pages 148–51) was the main group affected, with the elevation of 23 of its former subfamilies or tribes to family level: Boucekiidae, Ceidae, Cerocephalidae, Chalcedectidae, Cleonymidae, Coelocybidae, Diparidae, Epichrysomallidae, Eunotidae, Herbertiidae, Hetreulophidae, Heydeniidae, Idioporidae, Lyciscidae, Macromesidae, Melanosomellidae, Moranilidae, Neodiparidae, Ooderidae, Pelecinellidae (senior synonym of Leptofoeninae), Pirenidae, Spalangiidae, and Systasidae.

The family Calesidae was erected for the genus *Cales*, previously included in the Aphelinidae. Within the Eupelmidae the subfamily Neanastatinae was elevated to family rank, and the genus *Metapelma* was transferred to the newly erected family Metapelmatidae. Baeomorphidae was recognized as the senior synonym of Rotoitidae. The extinct family Khutelchalcididae was excluded from Chalcidoidea and its placement is now considered to be unknown within the Apocrita. This family is only represented by the Early Cretaceous fossil species, *Khutelchalcis gobiensis*.

The new families with their constituent genera are
listed here for completeness.

Boucekiidae
Boucekius, Chalcidiscelis

Calesidae
Cales

Ceidae
Bohpa, Cea, Spalangiopelta

Cerocephalidae
*Acerocephala, Cerocephala, Choetospilisca,
Gahanisca, Gnathophorisca, Laesthiola,
Muesebeckisia, Neocalosoter, Neosciatheras,
Paracerocephala, Paralaesthia, Sciatherellus,
Theocolax*

Chalcedectidae
Chalcedectus

Cleonymidae
*Agrilocida, Callocleonymus, Cleonymus,
Dasycleonymus, Notanisus, Zolotarewskya*

Coelocybidae
*Acoelocyba, Ambogaster, Ariasina, Coelocyba,
Coelocyboides, Cooloolana, Cybopella,
Erotolepsiella, Eucoelocybomyia, Fusiterga,
Lanthanomyia, Lelapsomorpha, Liepara,
Nerotolepsia, Ormyromorpha, Paratomicobia,
Tomicobomorphella*

Diparidae
*Cerodipara, Chimaerolelaps, Conodipara,
Conophorisca, Dipara, Diparisca, Dozodipara,
Hedqvistina, Lelaps, Myrmicolelaps, Netomocera*

Epichrysomallidae
*Acophila, Asycobia, Camarothorax,
Epichrysomalla, Eufroggattisca, Herodotia,
Josephiella, Lachaisea, Leeuweniella, Meselatus,
Neosycophila, Odontofroggatia, Parasycobia,
Sycobia, Sycobiomorphella Sycomacophila,
Sycophilodes, Sycophilomorpha, Sycotetra*

Eunotidae
*Cavitas, Cephaleta, Epicopterus, Eunotus,
Mesopeltita, Scutellista*

Herbertiidae
Exolabrum, Herbertia

Hetreulophidae
Hetreulophus, Omphalodipara, Zeala

Heydeniidae
Heydenia

Idioporidae
Idioporus

Lyciscidae (two subfamilies: Lyciscinae,
Solenurinae)
*Agamerion, Amazonisca, Chadwickia, Epistenia,
Eupelmophotismus, Grooca, Hadroepistenia,
Hedqvistia, Lycisca, Marxiana, Mesamotura,
Neboissia, Neoepistenia, Nepistenia, Paralycisca,
Parepistenia, Proglochin, Proshizonotus,
Protoepistenia, Riekisura, Romanisca,
Scaphepistenia, Shedoepistenia, Solenura,
Striatacanthus, Thaumasura, Urolycisca,
Westwoodiana*

Macromesidae
Macromesus

Melanosomellidae
*Aditrochus, Aeschylia, Alloderma,
Alyxiaphagus, Australicesa, Brachyscelidiphaga,
Encyrtocephalus, Epelatus, Espinosa,
Eurytomomma, Hansonita, Hubena, Indoclava,
Krivena, Lincolna, Lisseurytoma, Mayrellus,
Megamelanosoma, Nambouria, Neochalcissia,
Neoperilampus, Perilampella, Perilampomyia,
Plastobelyta, Queenslandia, Systolomorpha,
Terobiella, Trichilogaster, Westra, Wubina,
Xantheurytoma*

Metapelmatidae
Metapelma

Moranilidae (two subfamilies:
Moranilinae, Tomocerodinae)
*Amoturella, Aphobetus, Australeunotus,
Australurios, Eunotomyiia, Globonila, Hirtonila,
Ismaya, Kneva, Mnoonema, Moranila, Ophelosia
Tomicobiella, Tomicobomorpha, Tomocerodes*

Neanastatidae
Lambdobregma, Neanastatus

Neodiparidae (two subfamilies:
Elatoidinae, Neodiparinae)
Elatoides Neodipara

Ooderidae
Oodera

Pelecinellidae (two subfamilies:
Nefoeninae, Pelecinellinae)
Nefoenus, Doddifoenus, Leptofoenus

Pirenidae (five subfamilies: Cecidellinae,
Euryischiinae, Pireninae, Eriaporinae,
Tridyminae)
*Calyconotiscus, Cecidellis, Ecrizotes,
Ecrizotomorpha, Epiterobia, Eunotiscus,
Euryischia, Euryischomyia, Gastrancistrus,
Keesia, Lasallea, Macroglenes, Melancistrus,
Myiocnema, Oxyglypta, Papuaglenes, Petipirene,
Premiscogaster, Promuscidea, Sirovena,
Spathopus, Spinancistrus, Velepirene, Watshamia,
Zebe*

Spalangiidae (two subfamilies:
Erotolepsiinae, Spalangiinae)
*Balrogia, Erotolepsia, Eunotopsia, Papuopsia,
Playaspalangia, Spalangia*

Systasidae (two subfamilies: Systasinae,
Trisecodinae)
Semiotellus, Systasis, Trisecodes

LEFT | Head of *Trichilogaster
acaciaelongifoliae* (Melanosomellidae)
a biocontrol agent of introduced
invasive Australian wattle species
(*Acacia*, Mimosoidea) in South Africa.

SIGNIPHORID WASPS

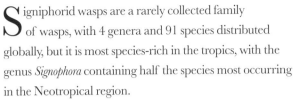

Signiphorid wasps are a rarely collected family of wasps, with 4 genera and 91 species distributed globally, but it is most species-rich in the tropics, with the genus *Signophora* containing half the species most occurring in the Neotropical region.

These wasps are either primary or hyperparasitoids attacking scale insects (Diaspididae, Pseudococcidae), mealybugs (Coccoidea), whiteflies (Aleyrodidae), and the hyperparasitoids via encyrtid or aphelinid wasp primary parasitoids. The obligate primary parasitoids attack fly (Diptera) pupae, or moth and butterfly (Lepidoptera) eggs. *Clytina* species and many *Signiphora* are primary parasitoids of Diptera puparia. The Indonesian *Signiphora bifasciata* is a parasitoid of the introduced Cycad Aulacaspis Scale (*Aulacaspis yasumatsui*, Diaspididae), and has potential as a biocontrol agent of this pest scale. Hyperparasitoid species have a negative impact on biocontrol programs, attacking useful primary parasitoid biocontrol species.

BELOW | *Signiphora* female, parasitoids or hyperparasitoids of aphids, psyllids, mealybugs, scale insects (Hemiptera), or flies (Diptera).

DISTRIBUTION
Worldwide

GENERA
Chartocerus, Clytina, Signiphora, Thysanus

HABITATS
A wide range, from forests to tundra

SIZE
$1/64$–$1/16$ in (0.2–1.5 mm) body length

ACTIVITY
Diurnal. Adults are rarely encountered, although can frequently be found by sweep netting

REPRODUCTION
Females lay eggs into their host insect

DIET
The larvae feed on the body contents of their host insect

TANAOSTIGMATID WASPS

This widespread family of rare wasps is concentrated in the tropics or subtropics, with highest species richness in the Neotropical region. The family contains 8 extant genera, with 103 species, and a single fossil genus.

Most species are phytophagous, forming galls in stems, leaves, seeds, or deformed ovaries of flowers. They mostly form galls on species of Fabaceae, but also on Euphorbiaceae, Lecythidaceae, Malvaceae, Myrtaceae, Polygonaceae, and Rhamnaceae. Some species develop as inquilines within galls formed by other insect species. Brazilian *Tanaostigmodes* species are inquilines in galls formed by gall midges (Cecidomyiidae). Indian *Tanaostigmodes* species are seed predators without gall formation. The African *Tanaostigmodes tamboticus* is a gregarious phytophagous gall-former on *Spirostachys africana* (Tamboti) (Euphorbiaceae). The Brazilian *Tanaostigmodes anamariae* develops in fruits of "Jacarandá-Do-Campo" (*Machaerium acutifolium*, Fabaceae), and *T. fernandesi* is an inquiline in galls of the gall midge *Anadiplosis* (Cecidomyiidae) formed on the same host. The Chinese species *T. puerariae* may be useful for biocontrol of Kudzu (*Pueraria lobate*), which is invasive in the USA, given that the wasp makes deleterious leaf galls on this plant, with up to 50 galls formed on a single leaf.

DISTRIBUTION
Worldwide, but absent from much of the northern hemisphere

GENERA
Enigmencyrtus, Liebeliella, Microprobolos, Minapis, Protanaostigma, Tanaoneura, Tanaostigma, Tanaostigmodes

HABITATS
Tropical or temperate forests, woodland, dry scrub

SIZE
$1/32$–$1/16$ in (1–2 mm) body length

ACTIVITY
Diurnal. Adults are rarely encountered

REPRODUCTION
Females lay eggs in the host plant tissue

DIET
Gregarious, phytophagous gall-formers, and seed predators

TETRACAMPID WASPS

A worldwide, small family of rare wasps with 14 genera and 44 species placed in 3 extant subfamilies: Mongolocampinae, Platynocheilinae, and Tetracampinae. There are two extinct genera, *Distylopus* and *Bouceklytus*.

They are either egg parasitoids, attacking beetles (Chrysomelidae), with *Dipriocampe* species attacking eggs of sawfly wasps (Diprionidae), or are larval parasitoids of flies (Agromyzidae). Species of *Foersterella* are internal parasitoids of *Cassida* beetle eggs (Chrysomelidae).

Dipriocampe diprioni was introduced to North America from Europe for the biological control of some sawfly pest species: Common Pine Sawfly (*Diprion pini*), Spruce Sawfly (*Gilpinia hercyniae*), and the European Pine Sawfly (*Neodiprion sertifer*).

BELOW | *Epiclerus panyas*, a parasitoid of many species of leaf-miner flies (Agromyzidae) in Europe.

LEFT | *Tanaostigmodes albiclavus* (FAR LEFT) and *Tanaostigmodes howardii* (LEFT), both gall-formers on *Acacia greggii* and *Mimosa biuncifera* (Fabaceae) in North America.

DISTRIBUTION
Worldwide, but absent from New Zealand

GENERA
Afrocampe, Cassidocida, Diplesiostigma, Dipriocampe, Electrocampe, Epiclerus, Eremocampe, Foersterella, Kilomotoia, Mongolocampe, Niticampe, Platyneurus, Platynocheilus, Tetracampe

HABITATS
A wide range, from forests to tundra

SIZE
$^1/_{32}$–$^1/_{16}$ in (1–2 mm) body length

ACTIVITY
Diurnal. Adults are rarely encountered

REPRODUCTION
Females lay their eggs directly into the host insect or egg

DIET
Larvae feed on the contents of the host insect egg they are developing in, or on the body contents of their host insect

157

TORYMID WASPS

DISTRIBUTION
Worldwide

GENERA
82 genera. Some of the more common
or representative include *Boucekinus,
Chalcimerus, Glyphomerus,
Mantiphaga, Microdontomerus,
Palachia, Palmon, Podagrion,
Pseudotorymus, Torymoides, Torymus*

HABITATS
A wide range, from forests to tundra

SIZE
$1/32$–$3/16$ in (1–5 mm) body length.
Some have exceptionally long
ovipositors

ACTIVITY
Diurnal. Adults are sometimes
commonly collected

Torymid wasps are a worldwide family with 82 genera and 900 species in 6 subfamilies: Chalcimerinae, Glyphomerinae, Microdontomerinae, Monodontomerinae, Podagrioninae, and Toryminae.

The family includes phytophagous species and many that attack gall-forming insects or are inquilines feeding both on the gall tissue and the gall inhabitant, such as species of Chalcimerinae, Microdontomerinae, and Toryminae, which are parasitoids of gall midges (Cecidomyiidae) and gall wasps (Cynipidae) and also attack fruit flies (Tephritidae) and plant lice (Psyllidae).

Most Monodontomerinae are external parasitoids of solitary wasp and bee larvae, or they can be internal parasitoids of eggs of bugs (Heteroptera), or of pupae of moths and butterflies or sawflies (Symphyta).

The Podagrioninae are specialized, attacking the oothecae (egg cases) of praying mantids (Mantidae).

LEFT | *Torymus*, a typical-looking torymid, many of which are phytophagous or inquilines, but others are parasitoids.

TOP RIGHT | *Podagrion* attacking mantis eggs in the ootheca by ovipositing through the foamy covering.

ABOVE RIGHT | *Chrysochalcissa* attacking leaf-footed bug eggs (Coreidae).

REPRODUCTION
Females lay eggs on or into their host larvae

DIET
The larvae feed on the body contents of their host insect and/or on plant gall tissue

Palachia and *Podagrion* species are commonly reared from Mantis egg cases. The female wasp lays her eggs into the mantid's ootheca, which is a hard foamy structure enveloping the eggs that ironically evolved for protection against predation and parasitism. This evolutionary arms race seesaws back and forth, with these torymids currently being specialist parasitoids of this resource.

Some species, such as in the genus *Ecdamua*, which attack solitary wasps nesting in holes in wood, have very long ovipositors, as much as five or six times their body length, which evolved to reach the host wasp larvae deep in the wood.

TRICHOGRAMMATID WASPS

This common worldwide family of tiny wasps has 97 genera with about 900 species. It contains some of the smallest insects, with a total adult length ranging from a paltry ¹/₆₄ in (0.2 mm) (e.g., *Megaphragma*) to ¹/₁₆ in (1.5 mm).

All trichogrammatids are parasitoids of eggs of other insects. They are often solitary, but many species are gregarious, with multiple individuals developing within a single host. They attack a wide range of host insects, but most commonly bugs (Hemiptera), moths and butterflies (Lepidoptera),

and beetles (Coleoptera). A few species exhibit phoresy, hitching rides on their host.

This wasp family is of economic and ecological importance, with many species used in biocontrol programs against insect pests of agricultural crops and forestry plantations.

Some species, particularly those in the genus *Trichogramma*, have a very broad host range, attacking any egg they find. *Lathromeroidea silvarum* is an aquatic species submerging to attack eggs of water beetles (Hydrophilidae and Dytiscidae).

DISTRIBUTION
Worldwide

GENERA
97 genera. Some of the more common or representative include *Aphelinoidea, Chaetostricha, Hayatia, Megaphragma, Oligosita, Trichogramma, Trichogrammatoidea, Ufensia, Uscana*

HABITATS
A wide range, from forests to tundra

SIZE
¹/₆₄–¹/₁₆ in (0.2–1.5 mm) body length

ACTIVITY
Diurnal. Adults are rarely encountered

REPRODUCTION
Females lay their eggs directly into the host insect egg

DIET
The larvae feed on the contents of their host insect egg they are developing in

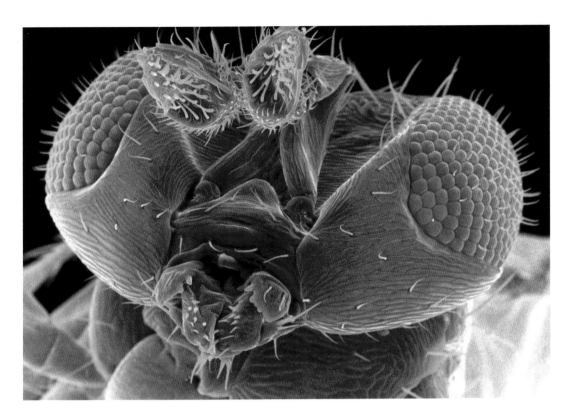

Megaphragma species have been studied as exemplars of the effects of extreme miniaturization on flying insects: the neurons of adults have mostly lost their nuclei and cells have merged, so the wasp has the bare minimum of cells to continue flying, feeding, and finding hosts, with the negative trade-off that life as an adult is very short.

ABOVE | Color-enhanced scanning electron micrograph of the head of a typical trichogrammatid taken at high magnification.

OPPOSITE | *Trichogramma evanescens*, an important biocontrol agent of pest caterpillar species attacking crops, here ovipositing into a moth egg.

RIGHT | *Trichogramma* female attacking an egg of the Angoumois Grain Moth (*Sitotroga cerealella*, Gelechiidae), a serious pest of stored grain.

BRACONID WASPS

Braconid wasps are second only to their sister group, Ichneumonidae, in species richness, with around 20,000 described species and many still undescribed. They comprise 1,057 genera placed in 41 subfamilies: Acampsohelconinae, Agathidinae, Alysiinae, Amicrocentrinae, Aphidiinae, Apozyginae, Brachistinae, Braconinae, Cardiochilinae, Cenocoeliinae, Charmontinae, Cheloninae, Dirrhopinae, Doryctinae, Euphorinae, Exothecinae, Helconinae, Homolobinae, Hormiinae, Ichneutinae, Khoikhoiinae, Macrocentrinae, Masoninae, Maxfischeriinae, Mendesellinae, Mesostoinae, Meteorideinae, Microgastrinae, Microtypinae, Miracinae, Opiinae, Orgilinae, Pambolinae, Proteropinae, Rhysipolinae, Rhyssalinae, Rogadinae, Sigalphinae, Telengaiinae, Trachypetinae, Xiphozelinae.

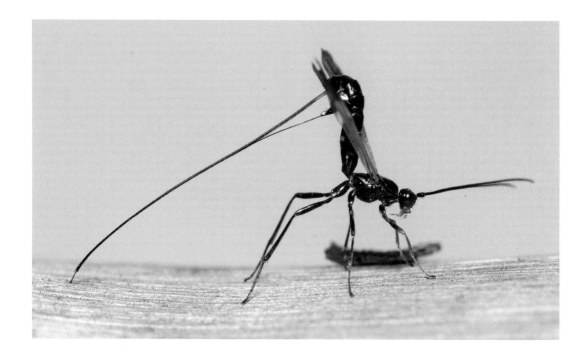

DISTRIBUTION
Worldwide, with the exception of the high Arctic and Antarctic regions

GENERA
1,057 genera. Some of the more common or representative include *Afrocampsis, Archibracon, Bathyaulax, Cardiochiles, Doryctes, Ecnomius, Gnamptodon, Helcon, Iphiaulax, Khoikhoia, Lysitermus, Macrocentrus, Opius, Orgilus,* *Pambolus, Phaenocarpa, Rhysipolis, Sigalphus, Thoracoplites, Vipio*

HABITATS
Wide range, from tropical forests and savanna ecosystems to cold, arid, temperate, and alpine environments

SIZE
Range in body size from $^1/_{32}$–$^9/_{16}$ in (1–15 mm) in length, some with extremely long ovipositors of up to $7^1/_8$ in (18 cm)

A huge variety of other insects are hosts to braconids, but Lepidoptera, followed by Diptera and Coleoptera, are the most frequent hosts.

Braconid subfamilies are grouped informally into the "cylostomes" (meaning that they usually have a distinct cavity above the mouthparts) and non-cyclostomes. The non-cyclostomes are all endoparasitoids (internally feeding) and koinobiont (allowing their hosts to develop further following oviposition), although there may be an external feeding phase toward the end of larval development. There is a much greater variety of life histories within the cyclostomes, including ectoparasitoids, endoparasitoids, idiobionts, and koinobionts. Two large subfamilies—Aphidiinae and Rogadinae—mummify their host larvae as hardened cadavers in which to pupate.

LEFT | *Cyanopterus* (Braconinae) female ovipositing onto a host beetle grub feeding inside wood.

TOP RIGHT | *Iphiaulax* (Braconinae), a parasitoid of wood-boring beetle larvae in the family Cerambycidae.

RIGHT | *Cotesia glomerata* (Microgastrinae) laying eggs into the first-instar caterpillars of the Large Cabbage White (*Pieris brassicae*, Pieridae).

ACTIVITY
Diurnal or nocturnal as adult wasps, larvae develop inside or on the host

REPRODUCTION
Females usually lay eggs into or onto hosts, with the larvae developing inside

DIET
Larvae feed on internal body tissues of host insects or, in a few cases, on plant gall tissue. Adults feed on nectar and sometimes on exuding host body fluids

While many braconids are small and inconspicuous, and typically smaller than Darwin wasps, there are numerous large and spectacular species, especially tropical species in the subfamilies Braconinae and Doryctinae, which are often part of Müllerian mimicry rings, with distinctive regional color patterns. The braconine *Euurobracon yokahamae* has an ovipositor measuring up to $7^{1}/_{8}$ in (18 cm), which is threaded through cracks to parasitize hosts deep in dead wood.

One subfamily, Euphorinae, are mostly specialist parasitoids of adult insects, or nymphs of hemimetabolous insects, attacking a range of groups such as bees, ants, lacewings, true bugs, bark lice, and crickets. These are internal parasitoids that emerge and spin their cocoons externally. One well-known euphorine is *Dinocampus coccinellae*, which is a parasitoid of various ladybird beetles (Coccinellidae); occasionally very large ladybird beetle hosts have stayed alive long enough after *Dinocampus* has emerged to then reproduce again.

Although usually strictly parasitoid, a feature of many of the non-cyclostome braconids is that the larvae feed on the host's hemolymph, leaving the body otherwise intact. The host usually still dies, probably due in large part to the large emergence holes made by the exiting larvae. Within the cyclostome subfamilies are some large groups of parasitoids of Diptera larvae (Alysiinae and Opiinae) and many parasitoids of a variety of concealed hosts. One doryctine genus, *Sericobracon*, is exceptional in parasitizing adult Embioptera (webspinners). There seem to have been several independent evolutionary reversals from a parasitoid lifestyle to one of plant-feeding (phytophagy) in the subfamilies Braconinae, Doryctinae, and Mesostoinae.

Many ichneumonoids are parasitoids of Lepidoptera, but perhaps the most successful radiation is the braconid subfamily Microgastrinae. Over 3,000 species have been described but the total species richness is estimated at between 30,000 and 50,000 species. One well-known microgastrine is *Cotesia glomerata*, a gregarious parasitoid of cabbage white butterflies (*Pieris* species, especially *Pieris brassicae*). *Cotesia glomerata* larvae emerge from well-grown caterpillars and spin a mass of yellow silken cocoons. The wasp larvae disrupt the hormones of the caterpillar, changing the host's behavior so that it walks up any available vertical surface and is quiescent while the wasp larvae emerge, before taking up position over the wasp cocoons and guarding them from predators, until the caterpillar dies.

Because of their often highly specialized parasitoid life histories, numerous braconids have been used successfully in biological control of crop pest insects, especially the subfamily Microgastrinae against lepidopteran pests, and the subfamily Aphidiinae against aphids.

LEFT | *Aphidius colemani* (Aphidiinae) female laying an egg directly into a live aphid. Her larva will consume the host's body contents and pupate inside the "mummified" skin of the aphid.

BELOW | Microgastrine cocoons spun on the outside of the host caterpillar wherein the larvae have developed as internal parasitoids.

DARWIN WASPS

Ichneumonidae, recently named Darwin wasps, as their parasitoid biology was given by Darwin as an argument against the idea of a beneficent creator, is one of the most successful of animal families, with over 25,000 species described from across the globe and perhaps 75,000 species awaiting description. They comprise approximately 1,600 genera in 41 subfamilies: Acaenitinae, Adelognathinae, Agriotypinae, Anomaloninae, Ateleutinae, Banchinae, Brachycyrtinae, Campopleginae, Claseinae, Collyriinae, Cremastinae, Cryptinae, Ctenopelmatinae, Cylloceriinae, Diacritinae, Diplazontinae, Eucerotinae, Hybrizontinae, Ichneumoninae, Labeninae, Lycorininae, Mesochorinae, Metopiinae, Microleptinae, Neorhacodinae, Nesomesochorinae, Ophioninae, Orthocentrinae, Orthopelmatinae, Oxytorinae, Pedunculinae, Phygadeuontinae, Pimplinae, Poemeniinae, Rhyssinae, Sisyrostolinae, Stilbopinae, Tatogastrinae, Tersilochinae, Tryphoninae, and Xoridinae. Five subfamilies—Labenopimplinae, Novichneumoinae, Palaeoichneumoninae,

DISTRIBUTION
Worldwide, with the exception of the high Arctic and Antarctic regions

GENERA
Approximately 1,600 genera. Some of the more common or representative include *Alomya, Anomalon, Ateleute, Brachycyrtus, Charops, Chriodes, Compsophorus, Crytea, Diaparsis, Diplazon, Enicospilus, Euceros, Goryphus, Ichneumon, Lycorina, Melanodolius, Mesochorus, Metopius,* *Netelia, Orthocentrus, Paraphylax, Perilissus, Phorotrophus, Pimpla, Poemenia, Pristomerus, Rhyssa, Syzeuctus, Xorides*

HABITATS
Associated with almost all terrestrial habitats, from tropical forests and savanna ecosystems to meadows, rivers, tundra, etc.

SIZE
On average $3/8$–$3/4$ in (10–20 mm) in length, but body size ranges from

ABOVE | *Netelia* (Tryphoninae), one of the nocturnally active, species-rich genera that are always pale-colored, the Ophioninae being another nocturnal subfamily of superficially similar-looking pale species.

LEFT | Male *Acroricnus stylator* (Cryptinae), a parasitoid of mud-dauber wasp larvae.

Pherombinae, and Townesitinae—are known only from fossils. Most subfamilies are arranged in three vast informal groupings: the pimpliformes, ichneumoniformes, and ophioniformes.

Darwin wasps have radiated as parasitoids of other insects, with larval and pupal Lepidoptera probably the most

around $^1/_{32}$ in (1 mm) to $2^{15}/_{16}$ in (75 mm) in length, and some species have an extremely long ovipositor of up to $3^{15}/_{16}$ in (10 cm).

ACTIVITY

Diurnal, crepuscular, or nocturnal as adult wasps, immature stages developing within the host insect or arachnid. Females spend most of their adult life searching for hosts, and males are usually found searching for newly emerged females to mate with

REPRODUCTION

Females lay eggs into (usually exposed) or onto (usually concealed) hosts, most commonly larvae or pupae of Lepidoptera, Coleoptera, Hymenoptera, and Diptera. The wasp larvae develop by feeding on the still-living eggs, larvae, pupae of insects, and arachnids from eggs to adults, resulting in the eventual death of the host. Pupation occurs inside or outside of the host remains

DIET

Majority of species are parasitoids, larvae feeding on body tissue of the host insect or arachnid. Larvae of some species are predatory, especially on spider egg sacs. Adults imbibe nectar, plant sap, or honeydew, and many species feed on hemolymph of insects to gather proteins needed for the development of their eggs

ABOVE | Suspended cocoon of Campopleginae (possibly *Casinaria*), a subfamily containing many species used in the biocontrol of pest moths, sawflies, and beetles.

OPPOSITE | The long-tailed giant Darwin wasps *Megarhyssa macrurus lunator* (left) and *Megarhyssa atrata lineata* drilling into a log to try and oviposit onto the wood-boring larvae of the Pigeon Horntail (*Tremex columba*, Siricidae), developing deep inside. The pale ovals are a membrane extension of the metasoma within which the ovipositor and sheaths are coiled up, to be released during the drilling process.

important hosts, followed by Hymenoptera, Diptera, Coleoptera, and a variety of smaller orders. Compared to their sister family, Braconidae, Darwin wasps have some clear differences in host range and ecology, particularly the importance of Hymenoptera as hosts, either sawflies or other parasitoids.

No Darwin wasps are parasitoids of hemimetabolous insects or of adult insects, but quite a number of species are parasitoids of spiders and their egg sacs. However, some *Mesochorus* species can be reared from true bugs as parasitoids of braconid wasps, and *Phygadeuon vexator* has been reared from earwigs (Dermaptera), from the host tachinid fly *Triarthria setipennis*, a parasitoid of earwigs.

Unusual host associations include *Obisiphaga stenoptera* being reared from pseudoscorpion egg sacs; *Agriotypus* species walking underwater to parasitize caddis (Trichoptera) pupae; several *Nemeritis* species attacking snake fly (Raphidioptera) larvae under bark of trees; and species of the subfamily Hybrizontinae being parasitoids of ant larvae.

Unlike many groups of insects, Darwin wasps seem to be about as species-rich in temperate regions as they are in the tropics, although least in arid areas, and different subfamilies predominate in different areas.

Darwin wasps develop internally (endoparasitoid) or externally (ectoparasitoid) in or on their hosts, permanently paralyzing the host (idiobiosis) or allowing the host to develop further (koinobiont).

There is a huge variety of life histories, which can only be sketched out here. *Pimpla rufipes* is a generalist endoparasitoid of moth and butterfly pupae, paralyzing the host and injecting anti-bacterial compounds to keep the host tissues fresh. Sometimes hosts of other orders are parasitized too. In contrast, *Stilbops vetulus* oviposits in the egg of one species of longhorn moth, *Adela reaumurella*. The wasp larva completes its development once the host has pupated in a case in the leaf litter.

Another egg-pupal parasitoid is *Collyria coxator*, a parasitoid of cephid sawflies, which has a serrated ovipositor for cutting through grass stems to reach the host eggs and has been introduced to North America for biological control of invasive stem sawflies. Other successful Darwin Wasp biocontrol species have generally been associated with forestry systems, including several species acting as important parasitoids of *Neodiprion sertifer*, the European Pine Sawfly, a major invasive pest in North America.

Other species could be considered to be potentially deleterious to natural control of pest insect species, such as the many species of *Mesochorus*, which are hyperparasitoids of other parasitoids, and the subfamily Diplazontinae, which are mainly parasitoids of aphid-feeding hoverflies (Syrphidae). The species-rich Mesochorinae and the small group Eucerotinae are all specialized hyperparasitoids and many Phygadeuontinae and Pimplinae utilize parasitoid cocoons once their hosts have finished feeding (pseudohyperparasitoids). The polysphinctines (subfamily Pimplinae) are parasitoids of active spiders and include some of the best-known examples of parasitoids altering their hosts' behavior, as they manipulate the spider's web-spinning.

Drilling rhyssines, and various other species, incorporate manganese in their ovipositor tip to harden the cuticle. Some Darwin wasps detect concealed hosts by vibrational sounding, tapping on the substrate and detecting the returning "echoes."

What we know of their ecology and biology is fairly limited compared to the huge numbers of species, and many surprises probably await.

APOCRITA
(ACULEATA)
STINGING PREDATORY
AND PROVISIONING WASPS,
BEES, AND ANTS

The aculeates, Aculeata, is a natural, monophyletic group of narrow-waisted wasps (Apocrita)—that is to say, they all share a common ancestor. As discussed earlier, we do not use "infraorder" here, as there are no other monophyletic infraorders in the Hymenoptera (see the phylogenetic tree on page 30). The nine superfamilies of aculeates include familiar stinging wasps, ants, and bees, as well as predatory groups such as the spider-hunting wasps, but also parasitoid groups such as the flat wasps. All the eusocial Hymenoptera are aculeates, with many Apoidea, Vespoidea, and some Pompiloidea constructing nests, which is one of the precursors to a truly social life history.

The aculeate wasps are defined by the possession of a stinger that delivers venom but is not used to lay eggs, although the stinging ability has been lost by some groups, most notably the stingless bees (Meliponini). The sting is used in defense and to paralyze prey, although in the bees the sting is exclusively used in defense. Some aculeates are renowned for their painful stings, such as some velvet ants (Mutillidae), spider-hunting wasps (Pompilidae), and ants (Formicidae).

LEFT | A cuckoo wasp (*Chrysis* species) at rest.

The venom of the social species is potent and painful to vertebrates, a necessary evolutionary development to protect a large nest resource from vertebrate predators, including humans. The evolution of venom is a critical step for the development of eusociality, which has evolved independently several times. Another key innovation was provisioning of larvae in a nest—essentially bringing food to the larvae rather than leaving larvae to develop on or in the food, as in parasitoids. Eusociality is the highest level of sociality with division of labor—see the Wasp Biology section (page 22) for more detail.

LEFT | Red-banded Sand Wasp (*Ammophila sabulosa*), protecting her nest from attack by a cuckoo wasp female *Hedychrum nobile* that is attempting to lay her egg within the nest of the sand wasp.

FLAT WASPS

This common, species-rich, worldwide family includes 2,340 species in 84 genera, and is classified in 5 extant subfamilies: Bethylinae, Epyrinae, Mesitiinae, Pristocerinae, Scleroderminae, and three extinct subfamilies.

Sexual dimorphism can be extreme, with many species having wingless females that resemble ants and males that are winged, and some species having both winged and wingless female forms. Most species are black or brown, but several South American species are metallic green.

Females sting and permanently paralyze their host, usually butterfly or moth caterpillars, or beetle larvae, either leaving them where they were found if in a concealed situation or dragging and hiding the paralyzed host in a crevice.

ABOVE | *Pristepyris*, predators of beetle larvae.

RIGHT | The wingless female of *Sclerodermus domesticus*, a worldwide species associated with human habitation as it preys upon beetle larvae, such as the Drugstore Beetle (*Stegobium paniceum*) and Common Furniture Beetle (*Anobium punctatum*) that are often found in dwellings.

DISTRIBUTION
Worldwide, but with highest species richness in the tropics

GENERA
84 genera. Some of the more common include *Bethylus*, *Cephalonomia*, *Epyris*, *Goniozus*, *Laelius*, *Pristocera*, *Pseudisobrachium*, *Sclerodermus*

HABITATS
From tropical or temperate forests, woodland, and dry scrub to alpine vegetation

SIZE
1/16–3/8 in (2–10 mm) body length

ACTIVITY
Diurnal or nocturnal

REPRODUCTION
Females of gregarious species lay many eggs on a single host species; solitary species lay a single egg

DIET
The larvae feed on the contents of their host insect's body

Some species exhibit maternal care, with the mother guarding her eggs and larvae, and in some cases females cooperate, laying eggs on a shared host and then working together to deter predators. This type of cooperative behavior has been of great interest in studies of the evolution of sociality in Hymenoptera, with some other bethylids intensively studied because of their biocontrol significance.

Pristocera rufa attacks the weevil *Pantorhytes szentivanyi* (Curculionidae), a serious pest of cacao plantations in Papua New Guinea. *Cephalonomia gallicola* is a gregarious parasitoid attacking the larvae and pupae of a serious stored-products pest, the Drugstore Beetle, *Stegobium paniceum* (Anobiidae).

ABOVE | Female *Odontepyris* tending her brood.

CUCKOO WASPS

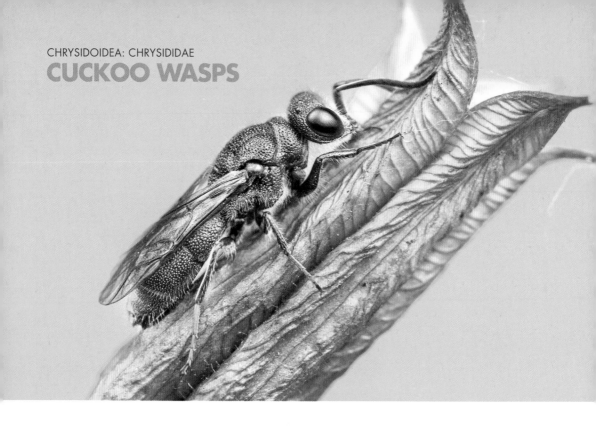

Cuckoo wasps are a species-rich worldwide family of strikingly colored wasps with 81 genera containing over 3,000 described species placed in 4 subfamilies: Amiseginae, Chrysidinae, Cleptinae, and Loboscelidiinae.

They are called "cuckoo wasps" because of their prevalent life history strategy whereby females of the most species-rich subfamily, Chrysidinae, lay eggs in other solitary wasp or bee nests and the cuckoo wasp larva consumes the host egg or larva and the food provision provided by the host wasp or bee female for the development of their offspring.

Praestochrysis shanghaiensis is an exception, being the only species known to attack caterpillars, developing on the final-instar, cocooned larva of the moth *Monema flavescens*. The female bites a hole in the hard cocoon to allow insertion of her ovipositor to sting and lay an egg onto the host caterpillar, and then reseals the hole.

The females in the subfamily Chrysidinae can fold into a compact ball to protect their more

DISTRIBUTION
Worldwide, but more common in arid areas

GENERA
81 genera. Some of the more common and species-rich include *Anachrysis*, *Chrysis*, *Cleptes*, *Hedychridium*, *Hedychrum*, *Loboscelidia*, *Praestochrysis*, *Stilbum*

HABITATS
Tropical or temperate forests, woodland, dry scrub, and deserts

SIZE
$^1/_{16}$–$^1/_2$ in (2–12 mm) body length

ACTIVITY
Diurnal. Adults are more common in hot, drier regions

REPRODUCTION
Females either lay an egg in the host wasp or bee nest, or into the host stick insect ootheca (egg sac), or on the host insect within a cocoon

DIET
The cuckoo kleptoparasitoid larvae feed on the host egg or larva plus the pollen and nectar food provision in the case of host bees, or insect prey provided as food by the host wasp for their larva. The egg parasitoids feed on the contents of the host egg. Adults take nectar

vulnerable undersides from the sting of the host wasp or bee whose nest they are parasitizing.

Amiseginae and Loboscelidiinae are parasitoids of stick insect (Phasmatodea) eggs within their oothecae. Cleptinae attack sawfly (Symphyta) larvae.

The cuckoo wasp family is also known as "emerald wasps" due to the brilliant metallic coloration of Chrysidinae species, which is often blue, green, or various shades of red or orange. Many species frequently have a combination of these colors in various patterns and configurations. These are reflective colors produced by the structure of the chitinous exoskeleton (see page 23 for more detail). The Amiseginae and Loboscelidiinae are non-metallic and usually brown or black.

Most species could be harmful in that they attack bees and wasps that perform pollination services, albeit at low density, but *Cleptes semiauratus* is a parasitoid of the Common Gooseberry Sawfly, *Nematus ribesii* (Tenthridinidae, page 52), a pest on red currant and gooseberry bushes.

PINCER WASPS

LEFT | *Dryinus spangleri* female with the visible pincer-like fore tarsi used to hold the host nymph during oviposition.

This diverse, worldwide family contains 53 genera and more than 1,800 described species in 16 subfamilies. Over half of the described species belong to three common genera: *Anteon*, *Dryinus*, and *Gonatopus*. They are parasitoids of leafhopper, planthopper, and treehopper nymphs and adults (Cicadellidae, Delphacidae, and Flatidae are most parasitized out of a total of 12 host families).

Females of some subfamilies are wingless and superficially resemble ants. They have a chelate fore tarsus for grasping the host hopper during egg-laying. This extraordinary feature is unique to dryinids and resembles a crab claw with rows of "teeth" along one "claw." The subfamily Aphelopinae differ in that females lack a chela.

The dryinid larva initially develops inside the host, but then extrudes in a characteristic hardened

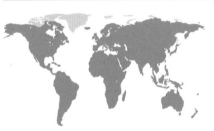

DISTRIBUTION
Worldwide

GENERA
53 genera. Some of the more common and species-rich include *Anteon*, *Aphelopus*, *Bocchus*, *Conganteon*, *Deinodryinus*, *Dryinus*, *Gonatopus*, *Neodryinus*, *Thaumatodryinus*

HABITATS
Tropical or temperate forests, woodland, dry scrub, arid areas

SIZE
$1/32$–$3/16$ in (1–5 mm) body length

ACTIVITY
Diurnal. Not commonly encountered, although the winged males can be abundant in insect traps

REPRODUCTION
Females lay an egg directly into the host insect with a needle-like ovipositor

DIET
Feed on body contents of their host insect

sac, the "thylacium," which bulges out of the host's abdomen. Pupation occurs on the host foodplant or in the soil within a spun cocoon.

Many species are considered to be important biocontrol agents of agricultural pests. They are not host-specific, often attacking a range of pest species. A single female *Gonatopus flavifemur* may attack hundreds of individual host hoppers to lay eggs into (up to 400), or to feed on (about 50 individuals) during her adult lifespan, which can last about 19 days. The North American *Neodryinus typhlocybae* effectively controls the Citrus Flatid Planthopper, *Metcalfa pruinose*, in Europe. The Chinese *Gonatopus nigricans* controls the Sugarcane Planthopper, *Perkinsiella saccharicida*, in Hawaii. The Rice Delphacid, *Tagosodes orizicolus*, is controlled by *Haplogonatopus hernandezae* in South America. *Echthrodelphax migratorius*, however, may have detrimental effects on biocontrol programs, since it attacks the South American Planthopper,

Megamelus scutellaris (Delphacidae), a useful biocontrol agent against the invasive Water Hyacinth (*Pontederia crassipes*).

INSET | A leafhopper (*Jikradia olitoria*, Cicadellidae) nymph parasitized by a dryinid larva. The larva first feeds inside the host, but then extrudes in a characteristic sac (seen on the underside of the host here).

BOTTOM | *Gonatopus taylori*, a wingless female that attacks leafhoppers (Cicadellidae).

179

EMBOLEMID WASPS

This is a rarely collected, small, worldwide family including 3 genera (*Ampulicomorpha*, *Embolemus*, and *Trogloembolemus*) and 62 species. Some authors consider *Ampulicomorpha* to be a synonym of *Embolemus*. This family also has 4 fossil genera, with the Cretaceous fossil genus *Baissobius* dating to approximately 140 MYA.

Females of *Embolemus* and *Trogloembolemus* are wingless or short-winged (brachypterous) and males are fully winged; both sexes of *Ampulicomorpha* are fully winged. However, the biology of this family is poorly known.

The North American *Ampulicomorpha confusa* was reared from planthopper nymphs (Achilidae) feeding

DISTRIBUTION
Worldwide

GENERA
Ampulicomorpha, Embolemus, Trogloembolemus

HABITATS
Tropical or temperate forests, woodland, dry scrub, fynbos

SIZE
$^{1}/_{16}$–$^{3}/_{16}$ in (2–5 mm) body length

ACTIVITY
Adults are rarely encountered

REPRODUCTION
In northern temperate regions females overwinter as adults

DIET
Poorly known. A Nearctic species was reared from planthopper nymphs (Achilidae) feeding on fungi under bark of decaying logs; other species attack planthopper nymphs in the Cixiidae

on fungi underneath bark of decaying logs; some species of Embolemidae have been reared from planthopper nymphs in the family Cixiidae. As in the pincer wasps, the larva develops in a bulging sac protruding from the intersegmental abdominal membranes of the host.

Females have been collected in ant nests and small mammal burrows. Embolemids may be ant mimics and several species are known to be associated with ants; for example, the Japanese *Embolemus walkeri* was collected from a *Myrmica* ant nest. The widespread Asian and European *E. ruddii* is associated with the ants *Formica fusca* and *Lasius flavus*, but probably attacks cixiid planthoppers feeding on tree roots.

Males of some species can be collected reasonably frequently in woodland, including *E. ruddii* in the autumn in Europe.

TOP | Female of the African species *Ampulicomorpha magna*.

ABOVE | Female of *Embolemus africanus*, which has highly reduced wings.

OPPOSITE | *Embolemus ruddii*, winged male, a parasitoid of tree root–feeding planthoppers (Cixiidae).

PLUMARIID WASPS

Containing 8 genera and 22 species, plumariid wasps are a rare and localized family occurring in the southern hemisphere. Four of these genera (*Maplurius*, *Mapluroides*, *Myrmecopterinella*, and *Pluroides*) are only known from a single species each. *Plumarius* is the most species-rich.

Sexual dimorphism is extreme. Females are wingless, with a flattened body and stout, spiny legs, suggesting that they are fossorial. They are rarely collected, usually from under rocks. Males have large wings and long legs and are commonly attracted to light or collected in Malaise traps in arid areas. Their biology is poorly known. The genus *Parapenesia*, which is only known from females, was originally placed in the Pristocerinae (Bethylidae).

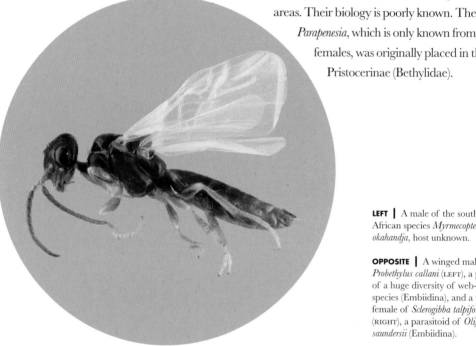

LEFT | A male of the southern African species *Myrmecopterinella okahandja*, host unknown.

OPPOSITE | A winged male of *Probethylus callani* (LEFT), a parasitoid of a huge diversity of web-spinner species (Embiidina), and a wingless female of *Sclerogibba talpiformis* (RIGHT), a parasitoid of *Oligotoma saundersii* (Embiidina).

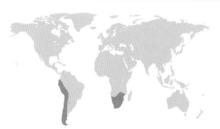

DISTRIBUTION
Arid and semi-arid areas in South America and southern Africa

GENERA
Maplurius, Mapluroides, Myrmecopterina, Myrmecopterinella, Parapenesia, Plumarius, Plumaroides, Pluroides

HABITATS
Arid habitats, including deserts, scrub, grassland, woodland

SIZE
$^1/_{16}$–$^1/_4$ in (2–7 mm) body length

ACTIVITY
Nocturnal. Females are rarely encountered, but males can be common at light traps

REPRODUCTION
Unknown

DIET
Unknown

SCLEROGIBBID WASPS

This family contains 3 genera represented by 21 species. They are external parasitoids of webspinner (Embioptera) nymphs or adults. Pupation takes place in the host web.

Sexual dimorphism is marked. Females are wingless, a likely evolutionary adaptation to facilitate negotiation of the host webspinner silk galleries, which can be spun on soil, rocks, or tree trunks, while males are fully winged and appear completely different. The sexes are impossible to associate without rearing or DNA analyses. Some fossil females are, however, fully winged.

Females insert only the first third of the egg into the host nymph so that it is mostly protruding; the larva that hatches then feeds externally. A parasitized webspinner can have up to 12 wasp larvae developing on it. *Probethylus schwarzi* attacks the webspinner *Anisembia texana* with the larva developing over a period of five to six weeks on the abdominal segments of the host nymphs, eventually killing and eating the remains of the host before pupating in white silken cocoons within the host web. The adults emerge two weeks later and actively negotiate the host web with ease.

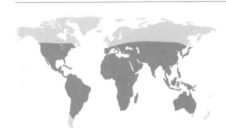

DISTRIBUTION
Worldwide, but absent from colder regions

GENERA
Caenosclerogibba, Probethylus, Sclerogibba

HABITATS
Tropical or temperate forests, woodland, dry scrub

SIZE
3/16–1/4 in (2.2–6.5 mm) body length

ACTIVITY
Diurnal. Females are rarely encountered, but males can be commonly collected in Malaise traps

REPRODUCTION
Females lay eggs partly into the host nymph

DIET
The larvae feed on the body contents of their host

SCOLEBYTHID WASPS

The scolebythid wasps are a rare family containing 4 genera and 6 extant species distributed mostly in the southern hemisphere. Three of these genera are represented by only a single extant species, and *Pristapenesia* by only two species.

These wasps are gregarious external parasitoids of wood-boring beetle larvae (Anobiidae, Cerambycidae, Ptinidae). *Pristapenesia stricta* females burrow through the frass-filled tunnels of their host beetle larvae, sting the beetle larva and feed on the hemolymph that exudes after biting through the exoskeleton, before laying five to seven eggs on the larva only a couple of days later. The wasp larvae feed externally on the host larva and pupate within the host burrow in wood.

Ycaploca evansi is present in both Australia and Africa, suggesting that this species may have been accidentally introduced to one or the other of these countries, probably facilitated by anthropogenic dispersal of infested wood containing the wasp larva developing on the host beetle. An *Ycaploca* has also been introduced to New Zealand.

RIGHT | *Ycaploca evansi*, a gregarious external parasitoid of wood-boring beetle grubs (Cerambycidae).

DISTRIBUTION
Australia, Brazil, Colombia, Costa Rica, Fiji, New Caledonia, New Zealand, southern Africa, Southeast Asia

GENERA
Clystopsenella, Pristapenesia, Scolebythus, Ycaploca

HABITATS
Tropical forests and woodland

SIZE
3/16–5/16 in (4–8 mm) body length

ACTIVITY
Diurnal. Adults are rarely encountered

REPRODUCTION
Females lay an egg on the host beetle larva developing deep inside branches or trunks after negotiating the tunnels that have been bored by the host larva

DIET
The larvae feed on the body contents of the host beetle larva. Adult females feed on the host hemolymph

ANTS

Ants are a hugely successful, worldwide, species-rich family with about 300 genera and 12,200 species. While species richness is not particularly high compared to some other Hymenoptera families, ant biomass is enormous, and they are vital components of nutrient and carbon cycling in most terrestrial ecosystems.

Ants are derived wasps, having evolved from within the aculeate wasp lineage. They are eusocial insects with distinct castes, the workers foraging, building the nest, protecting the colony, and feeding the brood, with only queens reproducing and males being produced at certain times of year.

ABOVE | The Indonesian golden ant *Polyrhachis ypsilon*, which at ⁹/₁₆ in (15 mm) is one of the largest of the more than 600 species in this Old World genus.

BELOW LEFT | The Yellow Crazy Ant (*Anoplolepis gracilipes*), a serious invasive species in many parts of the world, disrupting local ecosystems.

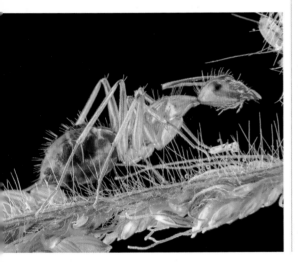

Some species have evolved to be social parasites or slave-makers, and these species have lost the worker caste. The queen may be replaced by laying workers, but these can only lay unfertilized, male eggs. Larvae are raised by the workers who feed them with animal, fungal, or plant matter.

The ecology of ants is varied, and their intricate behavior and social interactions have been the focus of many studies, which is outside the scope of this book. The ants are comprehensively dealt with in *Ants, A Visual Guide* by Heather Campbell and Benjamin Blanchard.

DISTRIBUTION
Worldwide

GENERA
300 genera in 16 extant subfamilies

HABITATS
From tropical forests to deserts and alpine areas

SIZE
¹/₃₂–2¹/₁₆ in (0.75–52 mm) body length

ACTIVITY
Diurnal or nocturnal

REPRODUCTION
Queens lay thousands of eggs over their lifespan of sometimes many years

DIET
Predatory, feeding on animal matter, or plant matter including fungus they cultivate

VELVET ANTS

LEFT | The Red Velvet ant (*Dasymutilla occidentalis*), also known as the Cow Killer because females can inflict an extremely painful sting in defense against predators, advertised by both their color and squeaking noise they make. They are parasitoids of large ground-nesting sand wasps (Bembicidae).

RIGHT | A winged male of *Ephutomorpha*, which looks completely different from the wingless females.

This is a common, worldwide family of solitary wasps with 216 genera represented by about 4,300 described species placed in 8 subfamilies: Dasylabrinae, Mutillinae, Myrmillinae, Odontomutillinae, Pseudophotopsidinae, Rhopalomutillinae, Sphaeropthalminae, and Ticoplinae.

Females are wingless and covered in velvet-like hairs, resembling large hairy ants, hence their common name. Males are usually fully winged and usually with different color patterns. Females have a potent and painful sting, a defense attribute that is advertised to predators using aposematic (warning) coloration, which can comprise various patterns of red, orange, yellow, or white on a black background, and via sound, as they produce a squeaky stridulation that warns potential predators of their harmfulness.

With their very hard cuticle and ability to roll into a ball, Mutillidae are difficult for potential predators (or bees and wasps trying to repel them) to harm. Combined with their stings, this means

DISTRIBUTION
Worldwide, but less diverse in temperate regions, and no native species are known from New Zealand

GENERA
216 genera. Some well-known genera include *Dasymutilla*, *Ephutomorpha*, *Mutilla*, *Pseudomethoca*, *Smicromyrme*, *Sphaeropthalma*, *Timulla*, *Traumatomutilla*

HABITATS
Tropical or temperate forests, woodland, dry scrub

SIZE
$3/16$–$15/16$ in (4–24 mm) body length

ACTIVITY
Most are diurnal but some are nocturnal

REPRODUCTION
The female seeks out hosts concealed in cervices or nests, commonly those

they have few enemies. Unsurprisingly, velvet ant species living in the same region have therefore often evolved similar color patterns, and these Müllerian mimicry rings help reduce sacrificial incidence during predator learning. They are also capable of running extremely fast when threatened. A few Australasian species are brilliant metallic blue, green, or purple. Sexual dimorphism is extreme, with gender association only possible through rearing, capture of pairs in copulation, or genetic analysis.

of ground-nesting wasps or bees, and lays a single egg near the host larva or pupa. Exceptions include some *Chrestomutilla* and *Smicromyrme*, which are parasitoids of the puparia of tsetse flies (*Glossina* species)

DIET
The larvae feed on body contents of the larva or pupa of other insects. A few species visit flowers, including copulating pairs of the subfamily Rhopalomutillinae

Velvet ants are external parasitoids of larvae or pupa of other insects, particularly the higher wasps, flies, butterflies and moths, beetles, and cockroaches. The larvae develop and pupate within the host nest, cocoon, or egg case. In several species phoretic copulation occurs, where the much larger winged male carries the smaller wingless female in flight during copulation.

A large North American species, *Dasymutilla occidentalis*, has earned the name Cow Killer because its sting is said to be so painful it could kill a cow. *Dasymutilla klugii* scored a 3 out of 4 on the Schmidt sting pain index, with the effects lasting for half an hour.

A few species living in open, arid environments have evolved camouflage rather than aposematic color patterns, with the best-known example being *D. gloriosa* in the southwestern United States, which closely resembles the seeds of the Creosote Bush (*Larrea tridentata*).

MYRMOSID WASPS

DISTRIBUTION
Asia, Europe, North America, and the
Oriental region

GENERA
*Carinomyrmosa, Erimyrmosa,
Krombeinella, Kudakrumia,
Leiomyrmosa, Myrmosa, Myrmosina,
Myrmosula, Nothomyrmosa,
Paramyrmosa, Pseudomyrmosa,
Taimyrmosa*

HABITATS
Temperate forests, woodland,
grassland, and deserts

SIZE
$^1/_4$–$^3/_8$ in (6–10 mm) body length

ACTIVITY
Diurnal

REPRODUCTION
The female seeks out larvae of
ground-nesting wasps or bees and lays
a single egg near the host larva or pupa

A northern hemisphere family, myrmosid wasps comprise 12 genera and 50 species. They were previously classified as a subfamily, or two subfamilies (Kudakrumiinae and Myrmosinae), in the Mutillidae; recent evolutionary studies have differed in their results, either classifying the myrmosids as the sister group to the (remaining) Mutillidae or as more closely related to sapygid wasps.

Sexual dimorphism is extreme, as it is in the velvet ants.

The winged males carry the wingless females during copulation in some species, but this is not the norm for this group.

They are external kleptoparasitoids of larvae or pupa of ground-nesting bees and wasps.

LEFT | A myrmosid, *Krombeinella longicollis*, a kleptoparasitoid of bee and wasp nests.

ABOVE RIGHT | *Myrmosula*, a North American genus attacking solitary bee and wasp nests.

RIGHT | *Myrmosa*, a Palearctic (European and Asian) genus kleptoparasite of ground-nesting bees.

DIET
The larvae feed on the body contents of the larva or pupa of other insects

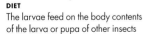

SPIDER-HUNTING WASPS

This common, worldwide family of solitary wasps contains 125 genera with about 4,860 described species placed in 4 subfamilies: Ceropalinae, Notocyphinae, Pepsinae, and Pompilinae.

Spider-hunting wasps are predators, or parasitoids, of spiders (Araneae), and occasionally other arachnids. Females attack living spiders and paralyze them using their sting, sometimes after a lengthy battle, and occasionally the spider wins. The paralyzed spider is then dragged to a predetermined nest site, usually a cell excavated by the female, but it may be a pre-existing cavity, where an egg is laid on the spider.

A single egg is laid on or in the abdomen of a spider that has been paralyzed by the sting of the female wasp, either in the spider's own burrow or on a spider that has been paralyzed, dragged, and placed in a secluded crack, crevice, excavated burrow, or mud nest made by the wasp. Some species are kleptoparasites of other species, laying their egg on a previously paralyzed and concealed spider. A few pompilids, such as the genus *Paracyphononyx*, are parasitoids of active spiders; the spider is only temporarily paralyzed, with the wasp larva developing on the active spider (as a koinobiont ectoparasitoid). Pompilids attack spiders from nearly all families, with genera and species

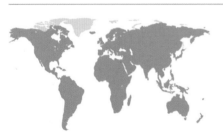

DISTRIBUTION
Worldwide

GENERA
125 genera. Some of the more common include *Anoplius, Auplopus, Batozonellus, Ceropales, Clavelia, Ctenocerus, Cyphononyx, Dichragenia, Hemipepsis, Irenangelus, Notocyphus, Pepsis, Pompilus, Pseudoclavelia, Tachypompilus*

HABITATS
From tropical forests to deserts and alpine regions

SIZE
3/8–2 3/8 in (10–60 mm) body length

ACTIVITY
Diurnal. Adults can be commonly encountered

tending to specialize on a related group of spider hosts. The larvae of Ceropalinae develop as kleptoparasitoids within another spider-hunting wasp's nest, or less frequently as an external parasitoid on a still-living spider.

Pepsinae include the large American tarantula hawks in the genus *Pepsis*, which have an extremely painful sting measuring 4 on the Schmidt sting pain index. This is almost at the top of the scale, with only the Bullet Ant, *Paraponera clavata*, measuring higher, at 4+. Most species in the subfamily Pompilinae are much smaller, although the Old World *Hemipepsis* species can be as impressive as *Pepsis*.

REPRODUCTION
Females lay an egg on
a paralyzed spider

DIET
Predators of spiders, the larvae
feeding on the body contents of
the paralyzed prey

ABOVE | The Australia Golden Spider Wasp (*Cryptocheilus australis*) hunts large spiders.

BELOW | The Leaden Spider Wasp (*Pompilus cinereus*) is widespread throughout the Old World.

CLUB-HORNED CUCKOO WASPS

These wasps are a small family of solitary parasitoid wasps with 12 genera and 66 species placed in 2 subfamilies: Fedtschenkiinae and Sapyginae.

Species of the sole genus *Fedtschenkia* in the subfamily Fedtschenkiinae occur in deserts or salt steppes in North America, Europe, and Asia and attack larvae of soil-nesting potter wasps (Eumeninae).

Sapyginae species occur worldwide with the exception of Australia. Adults have a black ground plan with yellow or white markings. They are kleptoparasitoids or external parasitoids of bee and wasp larvae (Megachilidae, Apidae [Anthophorinae], Eumeninae). Females lay eggs in the nests of solitary bees or wasps and the resultant hatching larvae consume both the host larva and the associated food provision.

In California *Fedtschenkia anthracina* was recorded as a parasitoid of the potter wasp *Pterocheilus trichogaster* that was provisioning its nest with looper caterpillars (Geometridae). The Egyptian *Sapyga luteomaculata* attacks the nests of the bees *Osmia submicans* and *Megachile minutissima* (Megachilidae), first killing and feeding on the host larva, before moving onto the stored provisions of nectar and pollen provided by the host mother bee for her own

ABOVE | *Sapyga nevadica* attacks nests of the leafcutter bee *Dianthidium dubium* (Megachilidae).

young. Some species are relatively easy to find as they attack solitary bees in artificial nests in gardens ("bee hotels"), such as *Sapyga quinquepunctata*, a parasitoid of Megachilidae bees in Europe.

Given that these wasps attack host species that are often useful pollinators of cultivated crops, they may have detrimental ecosystem impact. *Sapyga pumila* is a pest since it attacks the leafcutter bee *Megachile pacifica*, an important pollinator of Alfalfa crops in North America. *Polochrum repandum* is a large species attacking and developing on the carpenter bee *Xylocopa violacea*.

DISTRIBUTION
Worldwide except for Australia

GENERA
Araucania, Asmisapyga, Eusapyga, Fedtschenkia, Huarpea, Krombeinopyga, Monosapyga, Parasapyga, Polochridium, Polochrum, Sapyga, Sapygina

HABITATS
Tropical or temperate forests, woodland, dry scrub, fynbos

SIZE
$^3/_8$–$^9/_{16}$ in (10–15 mm) body length

ACTIVITY
Diurnal. Except where artificial nests are provided for aculeates, can be difficult to find

LEFT | The White-spotted Club-horned Wasp (*Sapyga quinquepunctata*) attacks nests of mason bees in the genus *Osmia* (Megachilidae).

BELOW | Larvae of *Monosapyga clavicornis* develop in nests of leafcutter bees (Megachilidae) made in dead wood or hollow stalks, feeding first on the host and then on its food provision of pollen and nectar.

REPRODUCTION
Females lay eggs in the nests of solitary bees or wasps

DIET
The larvae consume both the host larva and the associated food provision

BRADYNOBAENID WASPS

OPPOSITE | The southern African *Micatagla globularia*, apterous female (TOP) and winged male (INSET). Biology of the genus is still unknown.

BELOW | A Namibian species of *Apterogyna*.

B radynobaenids are a small family containing 5 genera and 236 species placed in 2 subfamilies, Apterogyninae and Bradynobaeninae, the former restricted to mostly arid areas of the Old World, although not found in Australia, and the latter found in Argentina and Chile. The New World Chyphoctinae and Typhoctinae have recently been transferred (as the family Chyphoctidae) to the superfamily Thynnoidea.

DISTRIBUTION
Much of the Old world, and also southern South America; basically found in arid areas

GENERA
Apterogyna, Bradynobaenus, Gynecaptera, Macroocula, Micatagla

HABITATS
Dry scrub, deserts

SIZE
$^3/_{16}$–$^9/_{16}$ in (5–15 mm) body length

ACTIVITY
Nocturnal. Adults are rarely encountered but may be commonly collected at light traps

REPRODUCTION
Unknown

DIET
Unknown

The black to reddish adults are usually hairy, occasionally with pale bands. Like the velvet ants, they have strong bodies with a thick exoskeleton and a powerful sting. Sexual dimorphism is extreme and gender association is reliant on capture of mating pairs or genetic analyses. Females move quickly on the ground and males can be found on flowers or at light. Nothing is known of the hosts of bradynobaenids.

MAMMOTH WASPS

T his large worldwide family of solitary wasps has 143 genera with about 560 species placed in 2 subfamilies, Proscoliinae and Scoliinae.

Mammoth wasps usually have a black ground plan marked with varying patterns of yellow, red, or white, and their wing tips have a distinctively corrugated appearance. Sexual dimorphism is less pronounced than in the Tiphiidae, with males slenderer than females.

DISTRIBUTION
Worldwide, although not found in colder areas, for example northwest Europe, and only present in New Zealand as an accidental introduction

GENERA
143 genera. Some of the more common or species-rich include *Campsomeriella, Campsomeris, Cathimeris, Colpa, Dielis, Megameris, Micromeriella, Scolia, Trielis, Triscolia*

HABITATS
Tropical or temperate forests, woodland, grassland, scrub, deserts

SIZE
$3/8$–$1^{15}/_{16}$ in (10–50 mm) body length

ACTIVITY
Diurnal

REPRODUCTION
Females burrow into the ground to lay an egg on the host beetle larva

DIET
The larvae feed on the body contents of their host beetle larva. Adults visit flowers for nectar and probably provide a pollination service

ABOVE | Male of the European Six-spotted Mammoth Wasp (*Colpa sexmaculata*). Males have characteristic upcurved spines at the end of the metasoma.

OPPOSITE | The Indonesian Giant Mammoth Wasp (*Megascolia procer javanensis*) has a wingspan of 4⅓ in (11 cm), making it one of the largest wasps in the world. It is a parasitoid of Atlas Beetle grubs (*Chalcosoma atlas*, Scarabaeidae).

Mammoth wasps are external parasitoids of beetle larvae (mostly Scarabaeidae) that live in soil or decaying vegetable matter, where pupation also occurs. Females burrow into the ground to locate the host to paralyze it by stinging. They need to feed on the hemolymph of their host larva before laying an egg on the host in its original burrow, or sometimes a new chamber is excavated nearby to receive the paralyzed host. Pupation takes place in a spun cocoon underground. Males may gregariously patrol areas, flying low to the ground in figure-of-eight patterns, waiting for females to emerge for mating opportunities.

These wasps are semi-useful biocontrol agents of pest beetle species, particularly white grubs attacking sugarcane, such as the Oriental *Exomala orientalis* accidentally introduced to the USA, which is controlled by *Micromeriella marginella modesta* [*Scolia manila*]. *Campsomeris* species introduced to USA to control the Japanese Beetle failed to effectively establish.

ABOVE LEFT | *Megascolia maculata flavifrons*, at 2³⁄₈ in (6 cm), is the largest wasp in Europe. It is a parasitoid of Rhinoceros Beetle grubs (*Oryctes nasicornis*, Scarabaeidae).

ABOVE | The Hairy Mammoth Wasp (*Scolia hirta*) commonly attacks Rose Chafer grubs (*Cetonia aurata*, Scarabaeidae).

The Giant Scoliid Wasp, *Megascolia procer*, is one of the largest wasps in the world, with many other species of stout proportions with wingspans of up to 2³⁄₈ in (60 mm), which in combination with their hairy appearance has led to the common name "mammoth wasps."

The South American orchid *Bipinnula penicillata* mimics a female scoliid, tricking males of *Pygodasis bistrimaculata* into pseudocopulation, resulting in an effective pollination event.

SIEROLOMORPHID WASPS

This rare, northern hemisphere family of solitary wasps has 2 genera and 13 species. *Proscleroderma* is represented by a single Syrian species, *Proscleroderma punctatum*, with the remaining 12 species belonging to *Sierolomorpha*. It has one extinct genus with only a single known fossil species.

Sierolomorphids are plainly colored black or brown species. Sexual dimorphism may be marked, with some species having wingless females. They are probably external parasitoids of other insects, but their biology is unknown.

ABOVE | *Sierolomorpha canadensis*, a rare species whose biology is unknown.

DISTRIBUTION
Central and Eastern Asia, North and Central America (Canada to Panama) and Hawaii

GENERA
Sierolomorpha, Proscleroderma

HABITATS
Tropical or temperate forests, woodland, grassland, alpine

SIZE
$^3/_{16}-^1/_4$ in (4–7 mm) body length

ACTIVITY
Diurnal

REPRODUCTION
Unknown

DIET
Unknown

TIPHIID WASPS

Tiphiid wasps are a large, worldwide family of solitary wasps in 24 genera with about 500 species placed in 2 subfamilies, Brachycistidinae and Tiphiinae. Brachycistidinae only occurs in North and Central America, while Tiphiinae are present worldwide.

Tiphiinae species are usually black, sometimes marked with red or yellow, while most Brachycistidinae are pale reddish, in common with many other nocturnal Hymenoptera. Sexual dimorphism varies from slight

DISTRIBUTION
Worldwide for the subfamily Tiphiinae, but Brachycistidinae are restricted to the Americas

GENERA
Acanthetropis, Brachycistellus, Brachycistina, Brachycistis, Brachymaya, Cabaraxa, Colocistis, Cyanotiphia, Dolichetropis, Epomidiopteron, Glyptacros, Hadrocistis, Icronatha, Krombeinia, Ludita, Mallochessa, Megatiphia, Neotiphia, Paratiphia, Paraquemaya, Pseudotiphia, Sedomaya, Stilbopogon, Tiphia

HABITATS
Tropical or temperate forests, woodland, dry scrub, deserts

SIZE
1/8–1 3/8 in (3–35 mm) body length

(for example, *Tiphia*) to extreme in brachycistidine species with wingless females. Males have a characteristic upcurved hook at the end of the abdomen.

They are external parasitoids of the larvae of scarab beetles (Scarabaeidae), darkling beetles (Tenebrionidae), and tiger beetles (Carabidae: Cicindelinae), in soil or decaying vegetable matter, where pupation also occurs.

The Oriental species *Tiphia vernalis* was introduced to the USA as a biocontrol agent of the invasive Japanese Beetle, *Popillia japonica*, which has a major impact damaging a variety of vegetable and fruit crops. *Tiphia vernalis* and *T. popilliavora* effectively control the white grub *Anomala orientalis*, accidentally introduced to the USA from Asia, which wreaks havoc in sugarcane plantations.

ACTIVITY
Tiphiinae are mainly diurnal, Brachycistidinae nocturnal

REPRODUCTION
Females burrow into the host substrate to find a larva to lay an egg on

DIET
The larvae feed on the body contents of their host beetle larva

CHYPHOTID WASPS

BELOW | Female *Chyphotes*; the winged males look very different.

This small family of New World solitary wasps contains 5 genera and 74 species in 2 subfamilies: Chyphotinae, with a single genus, *Chyphotes*, restricted to the Nearctic region with highest species richness centered in the southwestern United States; and Typhoctinae, ranging from the USA to northern South America (Typhoctes), with *Eotilla*, *Prototilla*, and *Typhoctoides* restricted to Argentina and Chile. Chyphotids can be common in deserts.

Prior to recent molecular studies, Chyphotidae was considered to comprise two subfamilies within the Bradynobaenidae. Even earlier, they had frequently been classified as velvet ants (Mutillidae).

Chyphotinae are nocturnal and Typhoctinae diurnal, but the biology of the family is poorly known. There is one record, without much detail, of a North American typhoctine species, *Typhoctes peculiaris*, as an external, solitary parasitoid of sun spiders (Arachnida: *Solifugae*), pupating within the host's burrow; this needs to be confirmed. Based on evolutionary relationships of this family, it has been postulated that the hosts are probably beetles in the families Carabidae, Tenebrionidae, or Scarabaeidae.

DISTRIBUTION
Argentina, Chile, southern Canada, Central America, and USA

GENERA
Chyphotes, Eotilla, Prototilla, Typhoctes, Typhoctoides

HABITATS
Arid scrub, deserts

SIZE
$^3/_{16}$–$^1/_2$ in (5–12 mm) body length

ACTIVITY
Diurnal or nocturnal

REPRODUCTION
Unknown

DIET
The larvae may feed on body contents of sun spiders, but more likely on beetle larvae

FLOWER WASPS

Flower wasps, a worldwide family, contains 52 genera and about 1,000 species placed in 5 subfamilies: Anthoboscinae, Diamminae, Methochinae, Myzininae, and Thynninae.

They are, as far as is known, mostly external parasitoids of beetle larvae. Many species parasitize Scarabaeidae developing in the ground or in decaying wood, pupation occurring in the host substrate. The Methochinae are specialists in attacking the predatory larvae of tiger beetles (Carabidae: Cicindellidae). Diamminae are an exception in that they are parasitoids of mole crickets (Orthoptera: Gryllotalpidae). The Australian *Diamma bicolor* attacks the mole cricket *Gryllotalpa coarctata*.

LEFT | A female of the Australian *Macrothynnus insignis*. The winged males of this species are fooled by the smell and appearance of the flowers of the Grand Spider Orchid (*Caladenia huegelii*) into thinking they are finding a female, and in the attempted copulation process provide a pollination service.

DISTRIBUTION
Worldwide, particularly species-rich in Australia and South America

GENERA
52 genera. Some of the more species-rich include *Agriomyia, Anthobosca, Diamma, Meria, Mesa, Methocha, Paramyzine, Thynnus, Zaspilothynnus*

HABITATS
Tropical or temperate forests, woodland, dry scrub, desert

SIZE
3/16–1 3/16 in (5–30 mm) body length

ACTIVITY
Diurnal

REPRODUCTION
The female stings each larva before laying an egg on it

DIET
Larvae feed on the body contents of the host, usually a beetle larva. Adults imbibe nectar, sap or host body fluids

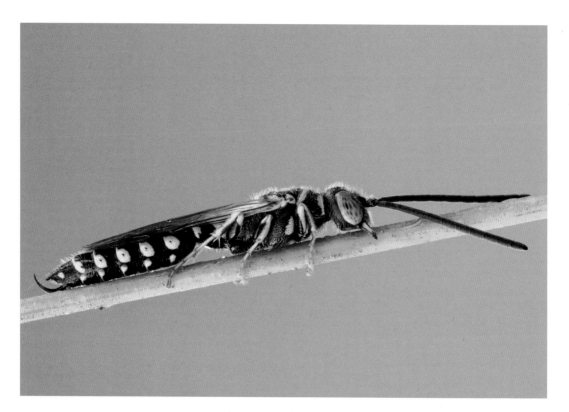

APOCRITA (ACULEATA)—Stinging Predatory and Provisioning Wasps, Bees, and Ants

Females of Diamminae, Methochinae, and Thynninae are wingless and smaller than the males. They are morphologically adapted to dig for their prey larvae, which they sting to paralyze before laying an egg on the host. In the remaining subfamilies both sexes are winged, and the males are black, often with white or yellow stripes and markings, and may be longer than the females, which are much stockier than the elongate males; a few species are metallic blue. The males usually have characteristic upcurved pseudo-sting projections at the end of the abdomen, which can be used defensively, although lacking venom.

Where the females are wingless, a pair mate in flight with the two wasps attached in copula, the male carrying the much smaller female for at least an hour.

A few species have been used for biological control.

Adults are flower visitors and feed on nectar and sap exuding from tree branches. Some females, however, only feed on the hemolymph of their host larva.

LEFT | *Meria tripunctata* attack Darkling Beetle grubs (Tenebrionidae) developing in soil.

BELOW LEFT | Female of the Australian *Diamma bicolor* commonly called a Blue Ant although that term is a misnomer. Females can deliver a painful sting and unusually for the family hunt mole crickets (Gryllotalpidae).

BELOW | The southern African *Paramyzine bidentata*, whose biology is unknown.

RHOPALOSOMATID WASPS

A widespread family, absent from Asia and Europe, Rhopalosomatid wasps comprise 4 extant genera and 72 species, in addition to 4 genera known only from fossils.

Rhopalosomatids are external parasitoids of cricket nymphs (Orthoptera: Gryllidae), the larvae forming extruded sac-like structures, rather like Dryinidae. Mature larvae pupate in a protected area.

BELOW | The Madagascan *Paniscomima rufoantennata*, the biology of which is unknown.

INSET | The very different southern African *Olixon saltator*.

DISTRIBUTION	SIZE
Predominantly tropical, found widely in the Americas, Africa, Madagascar, Southeast Asia, and Australia	$^3/_{16}$–$^9/_{16}$ in (4–15 mm) body length

DISTRIBUTION
Predominantly tropical, found widely in the Americas, Africa, Madagascar, Southeast Asia, and Australia

GENERA
Liosphex, Olixon, Paniscomima, Rhopalosoma

HABITATS
Tropical or temperate forests, woodland, dry scrub, deserts

SIZE
$^3/_{16}$–$^9/_{16}$ in (4–15 mm) body length

ACTIVITY
Diurnal or nocturnal

REPRODUCTION
Females sting and lay an egg on the host cricket

DIET
The larvae feed on the body contents of the host cricket nymph

A 100-million-year-old Cretaceous fossil in Burmese
amber is evidence that it is an ancient host association.
Individuals of the North American Restless Bush Cricket,
Hapithus agitator, can often be seen with rhopalosomatid larvae.

Species of *Olixon* look rather dissimilar to other
rhopalosomatids as they are small and ant-like, black or
reddish brown with extremely reduced wings, whereas the
fully winged *Paniscomima* and *Rhopalosoma* are large brown-
orange species that have very large eyes, both typical
characteristics of nocturnal wasps. Superficially these species
resemble some of the nocturnal Darwin wasps (Ophioninae,
Tryphoninae). The other Rhopalosomatinae genus, *Liosphex*,
is similar, but with darker coloration.

ABOVE | Rhopalosomatid larva
feeding externally on the Restless Bush
Cricket (*Hapithus agitator*, Gryllidae).

PAPER WASPS, POTTER WASPS, AND POLLEN WASPS

LEFT | The Asian Giant Hornet (*Vespa mandarinia*) builds its nest underground, which can contain thousands of individuals at seasonal peak. They can be a pest as they prey on honey bees.

RIGHT | The nest of the Common Wasp (*Vespula vulgaris*) constructed with a thick paper covering made up of chewed wood fibers and wasp saliva, enveloping the inner open-celled combs to help with temperature regulation.

BELOW | *Vespula vulgaris* queen.

This diverse, worldwide family includes 268 genera and about 5,000 species in 6 or 7 subfamilies: Eumeninae, Euparagiinae, Masarinae, Polistinae, Stenogastrinae, and Vespinae have been recognized for some time, with Zethinae sometimes elevated to subfamily rank rather than being classified within Eumeninae. Eumeninae, Euparagiinae, and Masarinae consist of solitary species, Polistinae and Vespinae are eusocial species, and Stenogastrinae contain both.

This family is represented by the best-known wasps because many of the social species have a painful sting that they use to defend their nests against vertebrate predators. The hornets, yellowjackets, and paper wasps are renowned in this regard.

DISTRIBUTION
Worldwide

GENERA
268 genera. A few of the more common include *Ancistrocerus, Belonogaster, Celonites, Delta, Eumenes, Jugurtia, Masarina, Odynerus, Polistes, Quartinia, Ropalidia, Rhynchium, Synagris, Vespa, Vespula, Zethus*

HABITATS
Very wide range of habitats, including tropical or temperate forests, woodland, dry scrub, deserts

SIZE
$^3/_{16}$–$1^{15}/_{16}$ in (5–50 mm) body length

ACTIVITY
Mostly diurnal; nocturnal genera include *Apoica* (Polistinae) and *Provespa* (Vespinae)

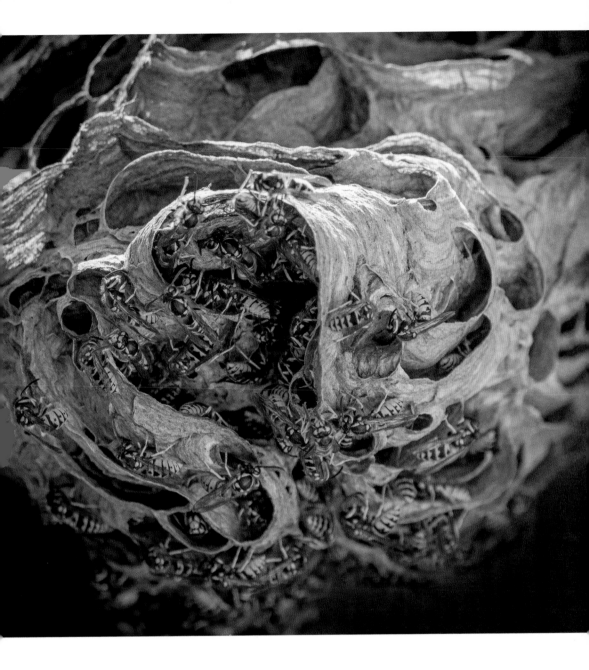

Paper wasps (including all of Polistinae and Vespinae) are social, constructing communal paper nests from chewed-up plant material and saliva. Larvae are fed on chewed-up insects such as caterpillars that are hunted by the females. However, active, predatory insects such as dragonflies (Odonata) can also be taken, and diets are usually varied and opportunistic, including scavenging on carrion.

The potter wasps (Eumeninae) are solitary and make their nests from mud or use existing cavities, provisioning the young with paralyzed caterpillars.

Pollen wasps (Masarinae) provision their nests with pollen and nectar and provide important ecosystem pollination services in arid areas.

The evolution of vespid social behavior, including the construction of their often highly intricate nests, has fascinated evolutionary biologists for a long time, along with that of ants and honey bees. The small tropical Asian subfamily Stenogastrinae (hover wasps) offer an interesting microcosm of social evolution, ranging from cooperative to eusocial societies, with Hover Wasps also using a variety of nest construction materials.

ABOVE LEFT | *Belonogaster*, a paper wasp female masticating a chewed-up moth caterpillar prior to feeding it to her offspring in a communal paper nest.

ABOVE | The African paper wasp *Polistes africanus*, which has fallen prey to a robber fly, *Hyperechia bifasciata* (Asilidae), which is a carpenter bee mimic. The little flies are jackal flies (Milichiidae) that hang around predators and scavenge on body juices that leak from their prey.

Some vespid species are invasive species in various parts of the world, introduced from their native Europe, such as the Common Wasp, *Vespula vulgaris*, and the German Yellowjacket, *Vespula germanica*, both species having a major impact on local ecosystem service providers in countries such as New Zealand and Australia. Some *Polistes* species, such as the European Paper Wasp, *Polistes dominula*, have become similarly problematic around the world. The Yellow-legged Hornet, *Vespa velutina*, native to Asia, is now established over much of Europe and is seen as one of several threats to the honey bee industry. Another specialized honey bee predator, the Asian Giant Hornet, *Vespa mandarinia* (the largest species of vespid at 1 15/16 in [50 mm]), has been accidentally introduced in North America. Just a handful of these hornets are capable of decimating 30,000 honey bees within a few hours.

Despite these numerous negative effects, the general contribution of vespid wasps is probably on balance economically positive, as the large colonies are sustained through eating insects, including many economic pests, and adults are undoubtedly pollinators.

COCKROACH WASPS

The nearly worldwide, although primarily tropical family of solitary predatory wasps nest in stems and crevices, comprising 6 genera and about 200 described species in the subfamilies Ampulicinae and Dolichurinae. *Ampulex* and *Dolichurus* are common genera found worldwide. *Aphelotoma* and *Riekefella* are restricted to Australia, *Trirogma* to Asia, and *Paradolichurus* to South America.

ABOVE | Emerald Cockroach Wasps (*Ampulex compressa*) hunt American Cockroaches (*Periplaneta americana*), stinging the host directly and very precisely in the central nervous system, subduing the prey into a compliant, living food supply for their larval development.

OPPOSITE | *Dolichurus corniculus*, a European and North African species that hunts cockroaches in the genus *Ectobius*.

DISTRIBUTION
Worldwide, but absent from colder regions, including New Zealand

GENERA
Ampulex, Aphelotoma, Dolichurus, Paradolichurus, Riekefella, Trirogma

HABITATS
Tropical or temperate forests, woodland, grassland, scrub

SIZE
$^3/_{16}$–$^9/_{16}$ in (4–14 mm) body length

ACTIVITY
Diurnal

REPRODUCTION
Females lay a single egg on the paralyzed host cockroach

DIET
The larvae feed on the body contents of the host cockroach. Adults feed on nectar

These wasps are all specialist predators of adult and nymphal cockroaches, which are weakly paralyzed and dragged backward by the wasp to a nest site in a concealed location, often led by an antenna. The female usually delivers a couple of stings, the first immobilizing the prey and the second targeting the part of the head ganglion that controls the escape reflex, effectively turning them into still-moving zombies, lacking any will of their own and becoming compliant. In that way the cockroach prey can be moved using its own powers of locomotion. The female wasp often nips the antennae of the cockroach and feeds on the exuding hemolymph. Adults will also take nectar at flowers.

The brightly colored metallic green or blue females (an attribute leading to the coining of alternative vernacular names for this family—the jewel or emerald wasps) may be seen scouting suitable habitats for prey, particularly old tree trunks, or beds of leaf litter. The nest is provisioned with only one cockroach. Usually, a single egg is laid on the mid-coxa of the host by the female and the nest is then closed with plant debris. When two eggs are laid, they both result in dwarf males.

Ampulex compressa attacks the ubiquitous, global pest cockroach *Periplaneta americana* and has potential as a biocontrol agent in this respect but may not utilize sufficient prey items to be effective at population control.

AMMOPLANID WASPS

Ammoplanid wasps are a worldwide family containing 10 genera and more than 130 species. Recent molecular studies produced the novel finding that Ammoplanidae are the closest relatives to the bees, the two lineages having diverged about 128 MYA.

They are solitary predatory wasps provisioning their nest with paralyzed thrips (Thysanoptera) for consumption by the larvae.

The American *Pulverro monticola* has been studied in detail and illustrates the ecology of ammoplanids. Females excavate burrows about $3^{1}/_{8}$– $3^{15}/_{16}$ in (8–10 cm) in length in earth banks. Their thrips prey (*Frankliniella moultoni* and *Aeolothrips fasciatus*) are carried to the nest in the female's mandibles and placed in several larger cells constructed at the bottom of the burrow. The nest entrance is not closed.

BELOW | An African *Ammoplanellus* female.

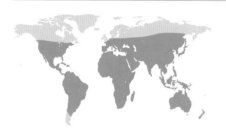

DISTRIBUTION
Worldwide, but absent from colder regions

GENERA
Ammoplanellus, Ammoplanops, Ammoplanus, Ammostigmus, Mohavena, Parammoplanus, Protostigmus, Pulverro, Riparena, Timberlakena

HABITATS
Tropical or temperate forests, woodland, dry scrub, deserts

SIZE
$^{1}/_{16}$–$^{1}/_{8}$ in (2–3 mm) body length

ACTIVITY
Diurnal

REPRODUCTION
Females provision their nest with host thrips before laying an egg in each cell

DIET
Larvae feed on the body contents of the host prey. Adults are frequent flower visitors taking nectar from various plants

APOIDEA: ASTATIDAE
ASTATID WASPS

BELOW | *Dryudella*, species of which prey on a range of at least eight true bug families (Hemiptera). Males have holoptic compound eyes (meeting on top of the head) that provide an excellent field of vision for finding females.

This worldwide wasp family contains 4 genera and more than 160 species. The genus *Astata* comprises about half of the known species.

Astatids usually prey on adults and nymphs of shield bugs (Pentatomidae), leaf-footed bugs (Coreidae), scentless plant bugs (Rhopalidae), and less commonly on crickets (Gryllidae), however they also attack a range of other true bug (Hemiptera) families.

A female of *Astata lugens* has been recorded carrying an immature cricket as prey in Brazil. *Dryudella stigma* provisions the single terminal cell in her nest burrow with about seven prey items before laying an egg. Burrows are excavated in sandy soil.

Adults visit flowers for their nectar requirements.

Males of *Astata* and *Dryudella* have large compound eyes that converge and meet on top of their head.

Astatid wasps are capable of extremely fast, albeit short, darting flights, executed from an elevated perch.

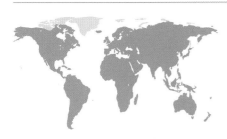

DISTRIBUTION
Worldwide

GENERA
Astata, Diploplectron, Dryudella, Uniplectron

HABITATS
Tropical or temperate forests, woodland, grassland, dry scrub, deserts

SIZE
3/16–5/8 in (4–16 mm) body length

ACTIVITY
Diurnal

REPRODUCTION
Females provision their nest with prey items before laying an egg in each cell

DIET
The larvae feed on the body contents of their host bugs or crickets. Adults visit flowers for nectar

SAND WASPS

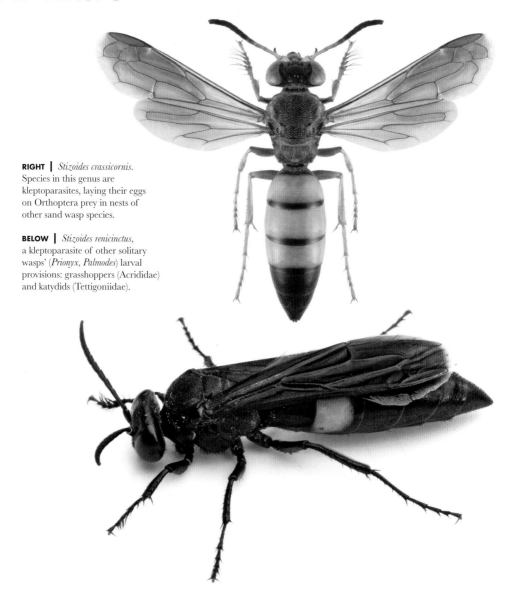

RIGHT | *Stizoides crassicornis.*
Species in this genus are
kleptoparasites, laying their eggs
on Orthoptera prey in nests of
other sand wasp species.

BELOW | *Stizoides renicinctus*,
a kleptoparasite of other solitary
wasps' (*Prionyx, Palmodes*) larval
provisions: grasshoppers (Acrididae)
and katydids (Tettigoniidae).

DISTRIBUTION
Worldwide

GENERA
81 genera. Some of the more common
include *Afrogorytes, Alysson,
Ammatomus, Bembecinus, Bembix,
Brachystegus, Didineis, Gorytes,
Handlirschia, Harpactus, Hoplisoides,
Hovanysson, Kohlia, Lestiphorus,
Nysson, Sphecius, Stizus, Stizoides*

HABITATS
Tropical or temperate forests,
woodland, grassland, dry scrub,
desert

SIZE
$3/16$–$1 9/16$ in (5–40 mm) body length

ACTIVITY
Diurnal

Sand wasps are a worldwide, common family of solitary wasps with 81 genera and 1,694 species in 3 subfamilies: Alyssontinae, Bembicinae, and Nyssoninae.

Females usually excavate nests in sandy soil and provision them with paralyzed prey for consumption by the larvae. Sand-rakes are present on the forelegs in many species, enabling efficient excavation of burrows.

A wide variety of insect orders is preyed upon. Most are predators but *Nysson* and *Stizoides* species are kleptoparasitoids of other sand wasps.

Sand wasps are capable of extremely fast and erratic flight. Most are small to medium-sized wasps, but the fearsome-looking (although very weakly stinging) cicada killers can be large (up to $1^{15}/_{16}$ in [50 mm] in length). They grab cicadas off tree branches, swooping to the ground before stinging the prey. *Sphecius speciosus* females carry paralyzed cicadas between their legs, which can be twice their weight, to a pre-excavated burrow before laying an egg on it. They provide a single cicada for a male egg but two to three cicadas for a female egg. The larger females need a bigger food resource for their development. These nests are often parasitized by the Cow Killer velvet ant *Dasymutilla occidentalis*.

Bembix species hunt flies, taking and paralyzing them while on the wing, with several species practicing progressive provisioning of prey items for their developing larva. They construct a multi-cellular sloping nest in sandy soil, some species nesting in aggregations of 1,000 or more individuals. *Bembix regnata* exceptionally preys on butterflies. *Bembix bubalus* females share a single nest and may practice division of labor.

Adults visit flowers, and some *Bembix* and *Stizus* species have been observed carrying pollinaria after visiting milkweed species, suggesting they provide a pollination service.

REPRODUCTION
Females provision their nests in sand with captured prey before laying an egg in each cell

DIET
The larvae feed on the body contents of the provisioned prey. Adults visit flowers for nectar

217

SQUARE-HEADED WASPS

Crabronidae are a diverse and large worldwide family of solitary wasps in over 110 genera and about 4,800 species placed in 2 subfamilies (Crabroninae and Dinetinae) and a series of tribes.

A wide variety of insect orders is preyed upon, often flies, with a few genera hunting unusual prey such as beetles, ants, or butterflies and moths.

They nest either in the ground, in hollow twigs, or in pre-existing cavities. A common species, *Ectemnius continuus*, makes nests in the soft pith of twigs. *Pison* and *Trypoxylon* females often build mud nests in sheltered situations or use mud to partition and seal cells in a nest made in pre-existing cavities, provisioning with paralyzed spiders. *Dasyproctus* females make their nests in plant stems and provision with flies. *Larra* females are unusual in that they hunt mole crickets (Gryllotalpidae), which are temporarily paralyzed for egg-laying after which the mole cricket recovers as a free-living host of the external parasitoid wasp larva. One species,

LEFT | *Ectemnius continuus*, a species widespread across much of the northern hemisphere. Females make nest burrows in dead wood and provision with flies (Syrphidae, Muscidae, Calliphoridae, Tabanidae).

RIGHT | *Oxybelus argentatus* female carrying her fly prey into her nest burrow excavated in the ground.

OPPOSITE BELOW | *Dasyproctus bipunctatus* emerging from her nest burrow made in a foxglove stem.

DISTRIBUTION
Worldwide

GENERA
Over 110 genera, the more common or representative including: *Belomicrus, Crabro, Crossocerus, Dasyproctus, Larra, Lestica, Liris, Miscophus, Nitela, Oxybelus, Palarus, Pison, Rhopalum, Tachysphex, Tachytes, Trypoxylon*

HABITATS
Tropical or temperate forests, woodland, dry scrub, fynbos

SIZE
$^1/_{16}$–$^{15}/_{16}$ in (2–24 mm) body length

ACTIVITY
Diurnal

REPRODUCTION
Females prey on a variety of insects as a food provision for their offspring

DIET
Larvae feed on prey items

Larra bicolor, is a biocontrol agent of pest mole crickets in the USA. Females of *Liris* hunt crickets (Gryllidae) but provision a nest, often in a pre-existing cavity, with the prey. Females of *Oxybelus* transport captured flies impaled on their sting, and the males defend the nest entrance. *Tachytes* and *Tachysphex* females hunt grasshoppers, katydids, or mantids.

In males of many species of *Crabro* the forelegs are modified into expanded, thin plates, which the male holds over the female's eyes during mating, a light-filtering process that may form part of mate recognition.

Adults commonly visit flowers for nectar.

ENTOMOSERICID WASPS

BELOW | *Entomosericus concinnus*, one of only three species in this family.

This is a rare, small family comprising a single genus, *Entomosericus*, with three species.

This rather aberrant genus has been classified in various ways, most recently as a tribe of Pemphredonidae (before the pemphredonids were raised to family status).

As the taxonomic position of *Entomosericus* has not been resolved in recent phylogenetic analyses, the genus is allocated to its own family here.

Females of *E. kaufmani* excavate almost vertical burrows in the ground and provision the nest site with immature and adult leafhoppers (Cicadellidae). The biology of the other two species, *E. concinnus* and *E. hauseri*, is unknown.

DISTRIBUTION
Central Asia to the eastern Mediterranean

GENUS
Entomosericus

HABITATS
Woodland, dry scrub, deserts

SIZE
$^1/_4$–$^5/_{16}$ in (6–8 mm) body length

ACTIVITY
Diurnal

REPRODUCTION
Females prey on leafhoppers to provision the nest

DIET
The larvae feed on the body contents of their host leafhopper prey

EREMIASPHECIID WASPS

This wasp family is rare and small, comprising a single genus, *Eremiasphecium*, with 11 species. It was previously placed as a tribe in the subfamily Philanthinae (now the family Philanthidae) within Crabronidae.

As this enigmatic genus is unplaced in recent work, it is treated as a family here.

Eremiasphecium species are solitary predatory wasps provisioning their nest with paralyzed thrips (Thysanoptera) for consumption by the larvae.

BELOW | *Eremiasphecium sahelense*, a North African solitary predatory wasp.

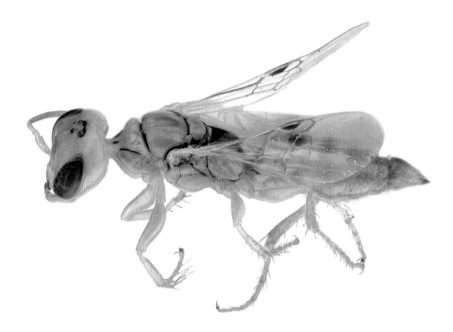

DISTRIBUTION
Central Asia to Northern Africa
(Canary Islands, Egypt, Kazakhstan,
Mauritania, Mongolia, Niger, Oman,
Russia, Saudi Arabia, Senegal,
Turkmenistan, United Arab Emirates,
Uzbekistan, Yemen)

GENUS
Eremiasphecium

HABITATS
Woodland, dry scrub, deserts

SIZE
$^{3}/_{16}$–$^{1}/_{4}$ in (4–6 mm) body length

ACTIVITY
Diurnal

REPRODUCTION
Females prey on thrips as a food
provision for their offspring

DIET
The larvae feed on the body contents
of their host prey

HETEROGYNAID WASPS

BELOW | The Madagascan
Heterogyna ravenala, the biology
of which is unknown.

This small family comprises a single genus, *Heterogyna*, with eight relatively tiny described species, and is very rarely collected.

Males are fully winged but females have reduced wings. The biology of these wasps is unknown.

Flightless females are unique to *Heterogyna* within the superfamily Apoidea, where females always transport food to their nest. Without the advantages of wings, it is more difficult for flightless females to transport prey, so it has been speculated that Heterogynaidae could be parasitoids, which would also be unique within Apoidea.

DISTRIBUTION
Afrotropical (Botswana, Madagascar) and Western Palearctic regions (Oman, Turkmenistan, United Arab Emirates, and eastern Mediterranean region)

GENUS
Heterogyna

HABITATS
Tropical forest (*H. madecassus*) or arid habitats

SIZE
1/16–3/16 in (1.5–5 mm) body length

ACTIVITY
Diurnal, except for the nocturnal *H. nocticola*

REPRODUCTION
Unknown

DIET
Unknown

APOCRITA (ACULEATA)—Stinging Predatory and Provisioning Wasps, Bees, and Ants

MELLINID WASPS

ellinid wasps are a mostly northern hemisphere (also occurring in South America) family with 2 genera and 17 species.

They are solitary predatory wasps, provisioning the nest with paralyzed flies (Diptera) for consumption by the larvae. Females mostly hunt around dung or tree leaves where flies gather, although *Mellinus crabroneus* is an exception, hunting around flowers. They stalk and jump on their stationary prey, grasping the wings between the mandibles so that they can sting the fly under the thorax.

They are gregarious nesters in open, sandy areas, each burrow having about ten cells at the bottom, which are provisioned with multiple flies before an egg is laid in the cell.

BELOW | Females of the Field Digger Wasp (*Mellinus arvensis*) construct nest burrows in sandy soil.

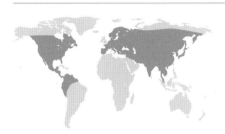

DISTRIBUTION
Nearctic, Neotropical, Oriental, and Palearctic regions

GENERA
Mellinus, Xenosphex

HABITATS
Tropical or temperate forests, woodland, grassland, dry scrub, desert

SIZE
$^3/_{16}$–$^3/_8$ in (4–10 mm) body length

ACTIVITY
Diurnal

REPRODUCTION
Females prey on flies as a food provision for their offspring

DIET
The larvae feed on the body contents of their host prey

APHID WASPS

T his diverse worldwide family of solitary wasps in 11 genera and about 1,000 species contains the smallest species (at $^1/_{16}$ in [2 mm]) in the superfamily.

Females provision their nests with Hemiptera (Homoptera), Thysanoptera, or Collembola.

Pemphredon and relatives exclusively prey on aphids, provisioning nests constructed either in the ground or in hollow twigs or pre-existing burrows in wood.

BELOW | The Mournful Wasp (*Pemphredon lugubris*), a widespread Palearctic species nesting in dead wood and provisioning with aphids.

DISTRIBUTION
Worldwide

GENERA
Arpactophilus, Carinostigmus, Diodontus, Microstigmus, Paracrabro, Passaloecus, Pemphredon, Polemistus, Spilomena, Stigmus, Xysma

HABITATS
Tropical or temperate forests, woodland, dry scrub, fynbos

SIZE
$^1/_{16}$–$^1/_2$ in (2–12 mm) body length

ACTIVITY
Diurnal

REPRODUCTION
Females prey on bugs, springtails, or thrips as food for their offspring

DIET
The larvae feed on the body contents of their host prey

Polemistus species commonly nest in the increasingly popular bee "hotels," where they are frequently attacked by torymid parasitoids in the genus *Ecdamua*.

Spilomena prey on thrips, or small bugs (Homoptera) such as aphids or coccids, carrying the prey with their mandibles to nests constructed in hollow twigs or rotten wood.

Stigmus and relatives prey on aphids, springtails, or thrips, are solitary or social, and may construct their own nests or use pre-existing cavities in other substrates.

Microstigmus are social and construct bag-like nests made from fibers or the waxy bloom of the plant from which they hang the nest (beneath a leaf). The cells are pockets made in the lower half of the nest while the adults live in the upper half of the bag.

ABOVE | *Passaloecus pictus* female using resin to close her nest made in a hole in a wall.

INSET | *Diodontus*, a northern hemisphere genus.

BEE WOLF WASPS

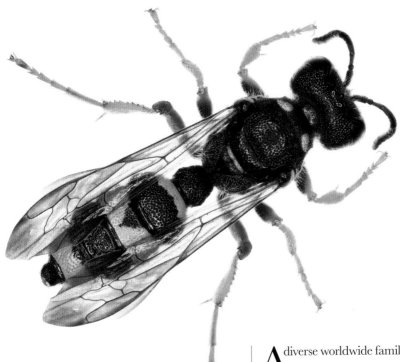

ABOVE | The Ornate Tailed Digger Wasp (*Cerceris rybyensis*) is a Palearctic species that hunts mining bees (Andrenidae).

OPPOSITE | The Beewolf (*Philanthus triangulum*) is a widespread species in Africa and the Palearctic region, which is a specialist hunter of the honey bee (*Apis mellifera*, Apidae).

A diverse worldwide family of solitary wasps, bee wolf wasps are classified in 8 genera and about 1,100 species. About three-quarters (around 900 species) of the described species belong to the genus *Cerceris*.

These are colorful wasps often marked with red or yellow stripes or spots. Most species of philanthids are medium-sized, but *Cerceris synagroides* is an exceptionally large species that is $1^3/_{16}$ in (30 mm) long.

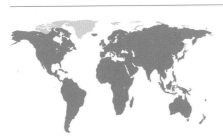

DISTRIBUTION
Worldwide

GENERA
Aphilanthops, Cerceris, Clypeadon, Eucerceris, Philanthinus, Philanthus, Pseudoscolia, Trachypus

HABITATS
Tropical or temperate forests, woodland, dry scrub, deserts

SIZE
$^1/_2$–$^{11}/_{16}$ in (12–18 mm) body length

on average, exceptionally up to $1^3/_{16}$ in (30 mm)

ACTIVITY
Diurnal

REPRODUCTION
Females prey on bees, ants, or beetles as food provision for their offspring

DIET
The larvae feed on the body contents of their host prey. Adults visit flowers for nectar

Many species prey on bees (e.g., *Philanthus*); *Clypeadon* prey on ants; *Cerceris* and relatives prey on bees, wasps, or beetles.

Females excavate burrows in the ground as nest sites and each cell is provisioned with several prey items. *Cerceris* species are usually host-prey-specific, but some such as *C. holconota* provision their cells with prey from as many as six families of wasps. The Ornate Tailed Digger Wasp, *C. rybyensis*, specializes on hunting small mining bees, which are usually taken when they are returning to their own nest with a pollen load—up to 30 individuals are sequentially captured to provision her nest with. *Philanthus* species are commonly known as

"bee wolves" due to their predilection for hunting bees as prey.

Cerceris nest aggregations are often characterized by the presence of surface "sand sausages", which are the accumulation of soil excavated during vertical burrow construction, the female pushing the excavated soil backward up the burrow with the hardened rear plate present on her abdomen (pygdium).

Adults are frequent visitors to flowers of a wide variety of families; they are commonly seen to be covered in pollen and most likely provide a pollination service.

APOIDEA: PSENIDAE

PSENID WASPS

Psenids are a small worldwide family of solitary wasps in 12 genera with approximately 460 described species.

Females prey on various Hemiptera but mostly on planthoppers (Cercopidae, Cicadellidae, Fulgoridae, Membracidae, and Psyllidae) as food for their offspring.

They nest either in burrows excavated in soil or rotten wood (such as *Mimesa* and *Psen*), and in that case these species usually have a rake present on the foreleg, or in pre-existing cavities such as in hollow or pithy plant stems, or holes in wood or mud banks, such as *Psenulus*, which lack a rake on the foreleg.

Mimumesa nigra females hunt leafhoppers to provision their nests excavated in rotten wood. *Psenulus* is the most species-rich genus in the family with over 160 species, centered in the Oriental region. *Psenulus* females are unique in that they line their nest cavity with silk produced from glands on their abdominal sternites.

Some species of *Pluto* nest in large aggregations in sandy areas.

Adults of psenids visit flowers to obtain nectar for their energy requirements. *Odontosphex damara* has been observed imbibing nectar from the extrafloral nectaries of *Euphorbia glanduligera* in Namibia. *Ammopsen masoni* is a frequent visitor to flowers of *Euphorbia* mats, such as the Matted Sandmat (*Euphorbia serpens*) and the Ground Spurge (*Euphorbia prostrata*), in the desert regions of the southern USA and Mexico.

ABOVE LEFT | *Psenulus fuscipennis* carrying a prey item to her nest entrance.

RIGHT | The Malaysian *Psenulus trimaculatus* in overnight roosting aggregation.

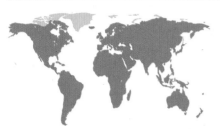

DISTRIBUTION
Worldwide, with most species occurring in the northern hemisphere

GENERA
Ammopsen, Deinomimesa, Lithium, Mimesa, Mimumesa, Nesomimesa, Odontopsen, Odontosphex, Pluto, Psen, Pseneo, Psenulus

HABITATS
Tropical or temperate forests, woodland, dry scrub, deserts

SIZE
$3/16$–$9/16$ in (4–15 mm) body length

ACTIVITY
Diurnal

REPRODUCTION
Females hunt planthoppers for nest provisioning before laying an egg in each cell

DIET
The larvae feed on the body contents of their host prey

APOCRITA (ACULEATA)—Stinging Predatory and Provisioning Wasps, Bees, and Ants

DIGGER WASPS, MUD-DAUBER WASPS, CATERPILLAR-HUNTER WASPS, THREAD-WAISTED WASPS

This diverse worldwide family of solitary wasps has 19 extant genera and about 800 species placed in 4 subfamilies: Ammophilinae, Chloriontinae, Sceliphrinae, and Sphecinae. There are also 2 fossil genera.

The abdomen is typically petiolate, which is to say there is a long, narrow "stalk." They make nests in pre-existing cavities, excavate burrows in soil, construct mud nests, or rarely make a nest from resin.

Females hunt and capture a diverse range of prey types, with each genus targeting specific prey types. Most provision their nest with multiple prey items before laying an egg in each cell. A few species of Sceliphrinae display basic sociality.

The mud-daubers in the genus *Sceliphron* construct mud nests on elevated surfaces and provision with spiders. The False Mud-dauber, *Chalybion spinolae*, hunts button spiders to provision

DISTRIBUTION
Worldwide

GENERA
Ammophila, Chalybion, Chilosphex, Chlorion, Dynatus, Eremnophila, Eremochares, Hoplammophila, Isodontia, Palmodes, Parapsammophila, Penepodium, Podalonia, Podium, Prionyx, Sceliphron, Sphex, Stangeella, Trigonopsis

HABITATS
Variety of habitats, from forests to tundra

SIZE
3/8–1 3/16 in (10–30 mm) body length

ACTIVITY
Diurnal

REPRODUCTION
Females capture and paralyze their prey before egg-laying, including cockroaches (Blattodea), crickets and

ABOVE LEFT | The Hairy Sand Wasp (*Podalonia hirsuta*) preys on large moth caterpillars (Noctuidae) to provision her nest, which is dug in soil after prey capture.

ABOVE | The Common Blue Mud-dauber of North America, *Chalybion californicum*, a specialist hunter of spiders.

grasshoppers (Orthoptera), caterpillars (Lepidoptera), larvae of other wasps, or adult spiders, as food provision for their offspring

DIET
The larvae feed on the body contents of their host prey

her nest, a short burrow that she makes in vertical banks. Nest-constructing *Chalybion* females use water to soften the hard mud for excavation, while other *Chalybion* species and females of *Isodontia* (which take long-horned grasshoppers as prey) use pre-existing cavities. *Chlorion* species take crickets, the New World species excavating nests, but the African species *Chlorion maxillosum* develops on large crickets (*Brachytrypes*) in the host's own burrow, the female wasp digging a tunnel to intersect with the host's burrow for attack from behind. The cricket is stung and paralyzed for egg-laying but recovers and digs a new burrow unaware of the presence of the wasp egg. *Prionyx* attack grasshoppers (Acrididae), whereas the large *Sphex* species hunt long-horned grasshoppers (Tettigoniidae) that can be twice or more their own body weight. *Ammophila* and *Podalonia* hunt caterpillars, and as do *Prionyx* and *Sphex* they provision nests made in friable soil.

A.

B.

Anthophila is a species-rich, conspicuous, worldwide group classified in seven families: Andrenidae, Apidae, Colletidae, Halictidae, Megachilidae, Melittidae, and Stenotritidae.

Bees are social wasps, having evolved from within the wasp evolutionary lineage, and are most closely related to the apoid wasps treated above.

They are distinguished by their usually hairy bodies with each hair branched unlike other wasp families, and in females a modified hind leg (or underside of the abdomen) evolved for collecting and carrying pollen, although this scopa is missing in socially parasitic species.

Most species are solitary and do not have a noticeable sting, making their nest in the ground, earth banks, pre-existing cavities made by other insects, or in hollow twigs.

Various grades of social organization are present across the seven families, culminating in the highly eusocial honey bee colonies.

DISTRIBUTION
Worldwide

FAMILIES
Andrenidae, Apidae, Colletidae, Halictidae, Megachilidae, Melittidae, Stenotritidae

HABITATS
Tropical or temperate forests, woodland, dry scrub, deserts

SIZE
$1/16$–$1^9/16$ in (2–40 mm) body length

ACTIVITY
Diurnal

REPRODUCTION
Varied, females lay an egg in a nest chamber or comb cell

DIET
The larvae feed on the pollen and nectar provision provided by their mother in the case of solitary bees, or by workers in the case of social species; some are parasitic on other bees

C.

D.

E.

F.

Bees are extremely important ecosystem service providers as pollinators of a huge range of flowering plants, including many agriculturally important species, a consequence of their foraging activities for nectar and pollen, protein sources that are used to raise their offspring.

The families of bees are comprehensively dealt with in a companion publication, *Bees of the World*, by Laurence Packer.

ABOVE | A. The European Tawny Mining Bee (*Andrena fulva*, Andrenidae), a solitary ground-nesting species that visits a wide range of flowers to obtain nectar and pollen for nest provision. Individual nests are often aggregated in suitable localities. **B.** Western Honey Bee (*Apis mellifera*, Apidae). **C.** The Australian *Stenotritus pubescens*, a fast-flying, sand-nesting bee, and member of the small family Stenotritidae. **D.** The widespread Palearctic species *Colletes cunicularius* (Colletidae), males of which pollinate sexually deceptive orchids (*Ophrys*) that mimick females. **E.** The Red Mason Bee (*Osmia bicornis*, Megachilidae), a solitary bee that nests in any suitable hole or hollow stem. **F.** The Texas Striped Sweat Bee (*Agapostemon texanus*, Halictidae), a ground-nesting species and generalist pollinator. **G.** The Pantaloon Bee (*Dasypodata hirtipes*, Melittidae), so named for the hairy hind legs of the females that expand in size with pollen collection.

G.

GLOSSARY

Arrhenotoky (Haplodiploidy): a sex-determination system where unfertilized eggs produce males (haploid) and fertilized eggs produce females (diploid).

Clade: group of evolutionary lineages sharing a common ancestor; clades are monophyletic.

Crepuscular: active at dawn and dusk.

Diploid: offspring containing two sets of chromosomes, one from each parent, because it develops from a fertilized egg.

Diurnal: active during the day.

Fynbos: a species-rich vegetation biome located in South Africa.

Haplodiploidy: see arrhenotoky.

Haploid: offspring containing one set of chromosomes from the mother, because it develops from an unfertilized egg.

Hemimetabolous: insect orders exhibiting incomplete metamorphosis: egg, nymph, adult.

Holometabolous: insect orders exhibiting complete metamorphosis: egg, larva, pupa, adult.

Hyperparasitoid: a species whose larva develops on a host parasitoid larva that is already attacking the primary host.

Idiobiont: a parasitoid that prevents further development of the host after oviposition. The hosts are usually developing in concealed situations. These parasitoids are often generalists.

Koinobiont: a parasitoid that allows the host to develop further following oviposition. The host is only partially paralyzed and recovers until killed on maturation of the parasitoid larva. These parasitoids are usually specialists.

Monophyletic: group of evolutionary lineages sharing a common ancestor.

Mycophagous: feeding on fungi.

Nocturnal: active during the night.

Oogenesis: process of egg maturation where the egg develops into a form capable of being fertilized.

Oviposition: the act of laying an egg.

Ovipositor: the egg-laying apparatus, which may be developed as a long "tail" projecting from the rear end of the female.

Paraphyletic: when two or more evolutionary lineages classified within a single taxonomic group share a common ancestor, but lineages classified separately are also included within the clade. For example, "fish," which share a common ancestor but some lineages of fish are more closely related to amphibians than they are to other fish.

Parasitoid: a species whose larva develops in (endoparasitoid) or on (ectoparasitoid) a host invertebrate that is still living or immobilized, eventually resulting in the death of the host.

Parthenogenesis: a form of reproduction that is asexual—progeny develop without fertilization, from an egg in animals, i.e., no fusion occurs between gametes (eggs and sperm).

Polyphyletic: when two or more evolutionary lineages classified within a single taxonomic group do not share a common ancestor.

Thelytoky: a reproductive subset of parthenogenesis where diploid females are produced from unfertilized eggs.

FURTHER READING

Blaimer, B. B., B. F. Santos, A. Cruaud, M. Gates, R. Kula, I. Mikó, J. Y. Rasplus, D. Smith, E. Talamas, S. Brady, and M. Buffington. 2023. "Key innovations and the diversification of Hymenoptera." *Nature Communications* 14:1212.

Branstetter, M. G., B. N. Danforth, J. P. Pitts, B. C. Faircloth, P. S. Ward, M. L. Buffington, M. W. Gates, R. R. Kula, and S. G. Brady. 2017. "Phylogenomic insights into the evolution of stinging wasps and the origins of ants and bees." *Current Biology* 27:1019–1025.

Branstetter M. G., A. K. Childers, D. Cox-Foster, K. R. Hopper, K. M. Kapheim, A. L. Toth, and K. C. Worley. 2018. "Genomes of the Hymenoptera." *Current Opinion in Insect Science* 25:65–75.

Burks, R., M.-D. Mitroiu, L. Fusu, J. M. Heraty, P. Janšta, S. Heydon, N. D.-S. Papilloud et al. 2022. "From hell's heart I stab at thee! A determined approach to rendering Pteromalidae (Hymenoptera) monophyletic." *Journal of Hymenoptera Research* 94:13–88.

Broad, G. R., M. R. Shaw, and M. G. Fitton. 2018. Ichneumonid Wasps (Hymenoptera: Ichneumonidae): their Classification and Biology (*Handbooks for the Identification of British Insects.* Vol 7, Part 12). Royal Entomological Society, St. Albans.

Chen, H.-Y., Z. Lahey, E. J. Talamas, A. A. Valerio, O. A. Popovici, L. Musetti, H. Klompen, A. Polaszek, L. Masner, A. D. Austin, and N. F. Johnson. 2021. "An integrated phylogenetic reassessment of the parasitoid superfamily Platygastroidea (Hymenoptera: Proctotrupomorpha) results in a revised familial classification." *Systematic Entomology* 46:1088–1113.

Eaton, E. R. 2021. *Wasps: The Astonishing Diversity of a Misunderstood Insect.* Princeton University Press, Princeton, New Jersey.

Evans, H. E. 1985. *Wasp Farm.* Cornell University Press, Ithaca, New York.

Goulet, H., and J. Huber. 1993. *Hymenoptera of the World: An Identification Guide to Families.* Agriculture Canada, Ottawa, Ontario.

Grissell, E. 2010. *Bees, Wasps, and Ants: The indispensable Role of Hymenoptera in Gardens.* Timber Press, Portland.

Heraty, J. M., R. A. Burks, A. Cruaud, G.A.P. Gibson, J. Liljeblad, J. Munro, J.-Y. Rasplus, G. Delvare, P. Janšta, A. Gumovsky, J. Huber, J. B. Woolley, L. Krogmann, S. Heydon, A. Polaszek, S. Schmidt, D. C. Darling, M. W. Gates, J. Mottern, E. Murray, A. Dal Molin, S. Triapitsyn, H. Baur, J. D. Pinto, S. van Noort, J. George, and M. Yoder. 2013. "A phylogenetic analysis of the megadiverse Chalcidoidea (Hymenoptera)." *Cladistics* 29: 466–542.

Malm, T., and T. Nyman. 2014. "Phylogeny of the symphytan grade of Hymenoptera: New pieces into the old jigsaw(fly) puzzle." *Cladistics.* 31:1–17.

Marshall, S. 2023. *Hymenoptera: The Natural History and Diversity of Wasps, Bees and Ants.* Firefly Books, Ontario, Canada.

Peters, R. S., L. Krogmann, C. Mayer, A. Donath, S. Gunkel, K. Meusemann, A. Kozlov et al. 2017. "Evolutionary history of the Hymenoptera." *Current Biology* 27:1013–1018.

Pilgrim, E. M., C. D. von Dohlen, and J. P. Pitts. 2008. "Molecular phylogenetics of Vespoidea indicate paraphyly of the superfamily and novel relationships of its component families and subfamilies." *Zoologica Scripta* 37:539–560.

Quicke, D.L.J. 2015. *The Braconid and Ichneumonid Parasitoid Wasps: Biology, Systematics, Evolution and Ecology.* John Wiley & Sons, New York. 704 pages.

Sann, M., O. Niehuis, R. Peters, C. Mayer, A. Kozlov, L. Podsiadlowski, S. Bank et al. 2018. "Phylogenomic analysis of Apoidea sheds new light on the sister group of bees." *BMC Evolutionary Biology* 18:71.

Sumner, S. 2022. *Endless Forms: The Secret World of Wasps.* HarperCollins, London.

van Noort, S. 2023. WaspWeb: Hymenoptera of the Afrotropical region. www.waspweb.org

Waldren, G. C., E. A. Sadler, E. A. Murray, S. Bossert, B. N. Danforth, and J. P. Pitts. 2023. "Phylogenomic inference of the higher classification of velvet ants (Hymenoptera: Mutillidae)." *Systematic Entomology* 1–25.

Evania appendigaster, page 74.

INDEX

PICTURE CREDITS

The publisher would like to thank the following for permission to reproduce copyright material (position on a page is indicated by L = left, R = right, T = top, B = bottom, M = middle):

Alamy Stock Photos 2020 Images 105T; B. Mete Uz 150B; BIOSPHOTO 99, 130, 133T, 146, 172–73, 201; blickwinkel 19, 131T, 219T; Charles Melton 156L, 156R; Clarence Holmes Wildlife 68, 73T, 73B, 101B, 108, 123, 127T, 135T, 147T, 148, 152; Daniel Borzynski 207; Denis Crawford 118, 161T, 164, 204B; Dorling Kindersley Ltd 212; Hakan Soderholm 213, 215; Henk Wallays 233BL; Larry Doherty 105B; LecartPhotos 17; Les Gibbon 163B; Maciej Olszewski 218L; Minden Pictures 117T; Mircea Costina 169; Natural History Collection 103; Nature Collection 233TL; Nature Picture Library 192, 193T; Nigel Cattlin 161B; Panther Media GmbH 191B; Phil Degginger 196; Richard Becker 223; Robert HENNO 198L; Tom Stack 16; Tomasz Klejdysz 40; Wirestock, Inc. 200. Alan Manson 69, 77, 85, 89, 116, 124, 141, 145T. Alex Wild 133 insert, 189T, 202. Arterra/Universal Images Group via Getty Images 232L. Maaminga marrisi. Auckland War Memorial Museum. AMNZ17769 © Auckland Museum 102. Ben Parslow 79, 82. Brian Valentine 111, 143. CC BY-SA 4.0_Bjørn Christian Tørrissen 194. CC BY-SA 4.0_Yasunori Koide 208T. Cor Zonneveld 220. Courtesy of the Trustees of the Natural History Museum 36, 46, 56, 86. Darren Ward 91. **Dreamstime** Henrikhl 43T; Macrero 76; Vladimir Zubkov 41. El Gritche 225B. Elijah Talamas, Florida State Collection of Arthropods 94; Ashton Smith 96; Melanie Anderson 95. Gavin Hazell 65. **Getty Images** Julie DeRoche/Design Pics 81; marcophotos 233MR; mikroman6 176. Henrik Gyurkovics 188. Ilona Loser 199. Jennifer Read 121B. Jeong Yoo 87. Jessica Joachim 180. John Heraty 121T. John Rosenfeld 80. Jonathan_Ball 163T. Matt Bertone 31, 48, 71, 122T, 122B, 147 insert, 174, 175T, 178. Matt Buffington 104, 113, 114, 115. Mayur Prag 27. Melvyn Yeo 100, 134, 135B, 151, 159T, 159B, 175B. Michele Menegon 12L. **Nature Picture Library** Alex Hyde 34–35, 52, 142, 149; Andy Sands 32, 43B, 193B; Doug Wechsler 42, 45TR, 45BL, 179T, 189B; Emanuele Biggi 185TR; Jiri Lochman 153T, 191T, 50, 51, 187, 203; Joao Burini 167; John Abbott 58, 59, 64, 74, 88L, 88R, 168, 216B, 234–35; Kim Taylor 18, 57, 140; Pierre Escoubas 139; Nick Upton 37, 166; Nigel Cattlin 24BL; Piotr Naskrecki 7B, 165, 137; PREMAPHOTOS 55; Rod Williams 53. Nicky Bay 75, 101T, 131B, 229. Paul Kitchener 144. Pekka Malinen 49, 90. Peter Webb 162. Pierre Bornand 70, 72, 84, 112, 119B, 120, 126, 127B, 128T, 128B, 157, 225T. Salvador Vitanza 138. **Science Photo Library** Heath Mcdonald 2, 197, 39, 204T. Sinclair Stammers 160; Stephen Ausmus/US Department of Agriculture 132; Steve Gschmeissner 119T. **Shutterstock** alslutsky 216T; Anton Kozyrev 208B; azrin_aziri 198R; Chase D'animulls 190; Dan4Earth 25; DeRebus 54T; gardenlife 217; Guenter Fischer/imageBROKER 125R; Henrik Larsson 44; Huw Penson 8TL; Hwall 224; IanRedding 106; irin-k 6; Kritchai7752 231; Kuttelvaserova Stuchelova 209; Lukman_M 185BL; Marco Maggesi 107T; Marieke Peche 232R; Melinda Fawver 5, 186; Muddy knees 54B; Reinier Blok 226; Sarah2 107B; South12th Photography 8BL; Timelynx 228; Tomasz Klejdysz 153 insert; Vitalii Hulai 230; Wirestock Creators 233TR, 233ML; Yunhyok Choi 177. Simon van Noort 7T, 12R, 13, 14 (composite of 48 images), 21B, 22TR, 22BL, 23, 24TR, 24ML, 26, 28, 29, 38, 66, 78, 92, 93, 97, 98, 117B, 125L, 129, 145B, 150T, 154, 155, 181T, 181B, 182, 183L, 183R, 184, 195, 195, 206, 206, 214, 221, 222. Steen Drozd Lund/Biosphoto/ardea.com 47. Stephen Marshall 20, 67. Vida Van Der Walt 8R, 9, 21T, 60, 62, 136, 158, 170, 179B, 205, 210, 211, 219BR, 227. Villu Soon/Natural History Museum, University of Tartu 83. Warren H L Wong 109, 110.

All reasonable efforts have been made to trace copyright holders and to obtain their permission for the use of copyright material. The publisher apologizes for any errors or omissions and will gratefully incorporate any corrections in future reprints if notified.

ACKNOWLEDGMENTS

Thanks to the world community of hymenopterists, both past and present, for their ongoing dedication and immense contribution toward exploring, unravelling, and documenting the world's amazing wasp, bee, and ant diversity and biology. We also thank the numerous professional and amateur photographers who provided images for use in this book, many of whom contribute valuable biodiversity records to citizen science platforms. Finally, we thank the editorial and production team at Bright Press for their unwavering dedication to the production of this volume on world wasps.